ROSS MARSHALL

MOON

MARS

MONUMENTS

MADNESS

"The Search for Alien Artifacts"

Volume - 2

By

R. S. Marshall

WeirdVideos.com

ROSS MARSHALL

Moon Mars Monuments Madness

"The Search for Alien Artifacts"

Volume - 2

2013 © Ross S. Marshall, Weirdvideos.com ISBN Number:

ISBN-13:978-1986008143

ISBN-10:1986008142

Cover Art credit by CreateSpace, Amazon Books

Printed in Anacortes, WA USA

All rights reserved. No part of this publication may be reproduced, distributed, or transmitted in any form or by any means, including photocopying, recording, or other electronic or mechanical methods, without the prior written permission of the publisher, except in the case of brief quotations embodied in critical reviews and certain other noncommercial uses permitted by copyright law. For permission requests, write to the publisher, addressed "Attention: Permissions Coordinator," at the address below.

Weirdvideos.com c/o R. S. Marshall

P. O. Box 1191

Anacortes, WA 98221

www.weirdvideos.com

Ordering Information: Quantity sales. Special discounts are available on quantity purchases by corporations, associations, and others. For details, contact the publisher at the addresses above.

Orders by U.S. trade bookstores and wholesalers. Please contact Weirdvideos Distribution: Tel: (360) 421-7195; or visit www.Weirdvideos.com

Printed in the United States of America

MOON

MARS

MONUMENTS

MADNESS
"The Search for Alien Artifacts"

Volume - 2

A Study in Selenography, whereby an attempt is made to determine whether there is evidence for extraterrestrial activity of space aliens. Illustrated with hundreds of NASA planetary pictures!

"When a person is honestly mistaken and hears the truth, they will either quit being mistaken, or they will cease to be honest." -- Anon

WeirdVideos.com

Dedication

This book is dedicated to George Leonard, Fred Steckling, Richard Hoagland, William Saunders, Mark Carlotto, Mike Bara, Rob Shelsky, George Kemplani, Erich von Daniken, Zecharia Sitchin, David Childress, Josephus Goodavage, Jack Swaney, Felix Bach, Don Ecker, Vito Saccheri, Rene Barnett, M. J. Craig, Victor Bertolaccini, Doug Turnbull, Tom Lehmann, Stephen Baxter, Ray Villard, Giorgio Tsoukalos, Stanley V. McDaniel and Monica Rix Paxson, Jason Martell, Philip Coppens, Louis Proud, Timothy Good, Alan McGregor, Stephen Webb, Maximillian de Lafayette, Whitley Strieber, Stanley McDaniel, Monica Paxson, Jim Marrs, Alex Milway, Adam Moon, Robin Moore, Xaviant Haze, Roc Hatfield, Mac Tonnies, George Haas, Brube Rux, Nick Redfern, Arthur C. Clarke, Gene Roddenberry and Lewis Carroll, with special thanks to Douglas Woodward.

CONTENTS

PREFACE   **1**
Did We Discover Alien Bases on the Moon?
Fred Steckling Sticks It to Us! A Refutation of
"We Discovered Alien Bases on the Moon"   29

1. Alien UFO Hangers, Ponds and Lunar Blobs   32
2. Moon Microbes   35
3. Irrigation Ponds   38
4. Clouds over Vitello   46
5. Alien Housing Projects and Crater Rim Domes   49
6. Fred's Damn'd Damoiseau Walls and Rille Rivers   54
7. Crater Cigars, UFO's and Airplanes   59
8. Double Craters and Listening Devices   65
9. Planetary Pies and Pie-cut Mounds   67
10. Crater Cabell's Alphabet Soup Letters   70
11. Tanks and Towers   75
12. Rolling Boulders and Mining Machines   80
13. Crater Lakes   91
14. Tsiolkowsky Water Crisis   95
15. Molehills, Mushrooms and Microbes   98

SECTION 4

JACK SWANEY'S "OBJECTS ON THE MOON"
Intermediate Idiotic Incidentals of Intergalactic Garbage

1. Mare Crisium Contraptions   101
2. More Alien Crisis on the North Shore   103
3. The Gigantic Block in Endymion   106
4. Robotic Bunnies in Crater Clavius   108
5. Thwe Rascally Rabbit in the East   110
6. Monumental Mountains of Moon Metal   111
7. Cranes and Booms of Julius Caesar   114

SECTION 5
MOON MARS MONUMENTS MADNESS - A POLEMIC DISCUSSION OF RICHARD HOAGLAND'S
"THE MONUMENTS OF MARS: A City on the Edge of Forever"   117

1	The Ukert Baseball Diamond	124
2	Return of the Ukertians	127
3	Dangling Crystals of Ukert City	130
4	A Glass Sphere Covering the Whole Moon?	133
5	Ukert Crater Triangle	135
6	Crystal Dome of Sinus Medii	138
7	Sinus Medii Dome Debunked	142
8	The Shard Location	148
9	The SHARD	153
10	The "Tower" and "Cube" Location	158
11	Where is the Shard?	163
12	Lunar Reconnaissance Orbiter Camera	171
13	Rebar Beams of Crater Manilius	175
14	The Castles of Ukert	182
15	The Mare Crisium Dome and Crystal Spire	192
16	More Ukert Crystals	200
17	NASA Cover-up or Touch-up?	202

FOOTNOTES   209
BIBLIOGRAPHY, ADVERTIZING   214

PREFACE

In the first volume *"Is Anyone Else on the Moon?"* we discussed George Leonard's supposed evidences for "Somebody Else Is on the Moon" and the alien presence theory. George was one of the first to popularize the theory (mythology) that aliens have visited our solar system and have occupied the moon and our other planets, starting a chain reaction that has evolved to great proportions. To do this, people like George Leonard must promote unjustified "conspiratorial" theories to pawn such nonsense off on the public at the expense of the credibility of NASA. This results in slandering and maligning the character of our space programs for hiding the so-called facts when there are no facts being hidden. This volume (volume-2) is the continued attempt in exposing the horrific problem of Fred Steckling and Richard Hoagland in seeing meaningful patterns or connections and objects (alien artifacts) in random or meaningless data.

Following the follies of Mr. Leonard comes Fred Steckling with his "We Discovered Alien Bases on the Moon" and Richard C. Hoagland with his best seller "The Monuments of Mars." Is it true that we are "not" alone in the Universe? Is there any truth to alien artifacts in backing up the claim of alien presence? We shall deal with this in the following volume.

The view that aliens inhabit our solar system and have left artifacts strewn about our planetary neighbors has spread like wild-fire into every corner of society and government. The alien presence "religion" is now a global reality and is considered true by more than half the world's population. It intrigues the rich, the poor, the educated and the illiterate as as well as politicians and religious people. In the last 60 years it has practically taken over the world, is universal, non-sectarian, indiscriminate and its membership exceeds that of all the religions of the world combined. In fact, most religions allow it to be believed because it is a religious belief itself with not a shred of evidence to back it up. So, what harm is it to believe?

The prevalence of this new "aliens are here" religion, and particularly the belief that life can be found on Mars is demonstrated with one internet poll revealing that 39.97% of the voters believed Mars is teaming with microbes. Another 40.85% (majority) supported Mars having had life a long time ago but not at the present, while 10.76% believed aliens live under the surface hidden from view, while 8.93% considered Mars stone dead, sterile and having no life now nor any in the past. Obviously, the poll taken showed almost 91% believing in some kind of extraterrestrial life at some time in the past, if not in the present.(1)

Everyone and their dog, from politicians to religious leaders seem to be playing party to the alien presence heresy. It is as if "aliens" are basically harmless, helpful and ultimately beneficial to humanity, albeit their obsessive compulsive habits of abduction and *palpatio per anum*. This must be true since they haven't destroyed us yet. They must have some vested interest in us. We must be valuable to some extent.

Little do people know of the "depth" of deception they are being sucked into. Like cattle to the slaughter, mankind is complacently falling head over heals into gleefully accepting the alien presence religion, which is leading to a complete ruination of our traditional foundations. Our foundations may be littered more than the moon has supposed artifacts with trashy history, but we are still alive today! Just because our past is filled with sin, corruption and error does not mean it is expendable and replaceable. At least we know what we have, where we've been and what lessons we have learned from it. Whereas, with aliens, we really have no idea what we are getting. By the fact that aliens are so illusive and because of the unjust abductions and experiments we have suffered, it should tell us something about the comfort we have gained from our hard earned past, and how much distrust we should have for this new paradigm from outer space. It may not be very logical, but it may very well be true, that by displacing the very traditional foundations of religions we do have, we open the door to our very destruction. It seems that the more we dispose of our traditions the more they have have influence. This tells us much about them. For example: Within all the UFO literature on our book shelves, is there one account where we may read that aliens have given us any credence for our hard earned religious beliefs?

PROPONENTS OF "WE ARE NOT ALONE"

Mr. Leonard greatly contributed to the modern belief that extraterrestrial physical life exists. Nevertheless, his arguments from morphological shapes, extreme highlights and shadows, and creative interpretations of surface features just do not hold up in scientific court rooms. Peer pressure needs hard evidence and George did not bring it to the table. Skeptics complain that all we see is a play of shadow and light upon funny shaped surface features, and that there are no such evidences. We saw this geophysical farce demonstrated in the first volume of "Is Anyone Else on the Moon? -- The Search for Alien Artifacts."

Despite the heavy skepticism, Fred Steckling, Jack Swaney, Richard Hoagland, and many others, continue promoting the Leonardean fiction of morphological madness. Aliens on the moon, lunar bases, extraterrestrial vegetation and other exotic signs of E.T. life are pushed to great extremes. Gene Roddenberry could not have done better. Skepticism has reclined if not retired and very few take it serious now.

The most prevalent contender for the existence of interstellar invaders is Mr Richard C. Hoagland. He has pushed the E.T. garbage to such an extreme that even government agencies, such as the National Aeronautics and Space Administration (NASA) have been persuaded to re-photograph supposed alien structures such as the Face on Mars. Many lunar and Martian features, claimed to be of extraterrestrial origin were re-photographed, scanned and analyzed along with the run-of-the-mill surface features. The desperate fetish to find fossils and artifacts in the photographic data sent back by our spacecraft has continued until every internet site, blog and forum, including NASA web sites, promotes moon and Martian artifact hunting.

ASTROBIOLOGY, LEAVEN IN THE NASA LUMP

Without much criticism from right-wing traditionalists (and for good lucrative reasons too - most of them have jumped in on the band wagon), astrobiology (and its derivative astropalaeobiology) has grown into the largest pseudo-religious cult on the face of the planet, and it is rapidly spreading to the moon, Mars and to all the other planets of our solar system. Traditional religious beliefs are diminishing in favor of the new belief that man is not alone and that one day it will be proved life exists beyond earth. There will then be no need for the prophets of old or the old middle-eastern Savior -- humanity will have found a new one - even if it is only a microbe.

LACK OF ASTROBIOLOGICAL EVIDENCE

It is a given thing today that government figures and a large amount of scientists think aliens exist (or at least extraterrestrial microbes) and that one day it will be proved they inhabit our solar system. Almost every space probe mission houses alien life detection equipment, despite the fact that for decades now and after many planetary missions, not one proof has been found.

Unfortunately for them the evidence so far is zero. The budgets are now being cut in favor of more reasonable and simplistic scientific experiments. Planetary probes are not looking for gigantic exo-paleontological evidences such as large alien structures, astrodomes, ruins of buildings or UFO airports. They are instead poking expensive microbial drills and other tools into the dirt, rocks and whatever else that might get in the way looking for biosignatures.

Over half the planet watches the incoming planetary data in hopes of finding alien artifacts, a piece of extraterrestrial junk or maybe catching an alien passer-by dropping their half-eaten jelly donut. Teams of scientists and engineers are now proposing a 2020 Mars rover mission that will borrow the design of the 1-ton Curiosity and seek out biosignatures on Mars (2).

One civilian enthusiast is actually sewing NASA for hiding supposed facts about what appears to him to be a jelly-donut or at least a fast growing Martian mushroom type plant - some say it is a humongous lichen. Others argue that a Martian must have past by and dropped the donut-looking "thing" either by mistake or on purpose, maybe to just taunt the human's space probe (3). The traditional NASA conservatives say it is a rock dislodged and flipped over by one of the wheels of the Rover. Could it be that a Martian tossed the sulfuric mushroom at the little alien rover thinking it might be hungry?

In the following study of alien artifacts we will look at Fred Steckling's "We Discovered Alien Bases on the Moon" and Jack Swaney's "Objects on the Moon" along with a fun and polemical exposé of Richard Hoagland's "The Monuments of Mars."

Our conclusion will be the same as volume one that both the moon and Mars are "stone" dead and adverse to any life forms; that they cannot naturally maintain life nor at anytime in the past have ever supported life forms. In a greater scope of cosmology, such dire lack of evidence infers greatly that even the universe itself is sterile and void of life, being just as adverse to life forms as our closest planetary neighbors - Mars, Mercury, Venus, Jupiter, Saturn and all the moons, etc. But, before we jump off into the alien artifact collection of imaginative artifices, let us look first at the evidence against extraterrestrial aliens and their habitats as existing outside the earth's atmosphere.

TWO OPPOSING PARADIGMS

Contemporary views upon the subject of the origin of life can be divided into two evolutionary branches of belief - Monogenetic Evolution (M.E.) and Polygenetic Evolution (P.E.) with the pro-life exobiologists on one side and the traditional geo-physicists and cartographers on the other.

Monogenists or single origin Evolutionists believe life and matter began from a single point or location and spread out from there. In other words, both matter and life had a big bang, where the former gave birth to the latter.

Alternately, polygenists or multiple origin Evolutionists believe that life popped up universally in many locations throughout the Universe, disconnected from one another and totally antonymous, where life is unique and completely independent in origins. Therefore, E.T. life must exist just as it does here on earth and should not have the same genetics.

M.E. postulates that all genetics ultimately originate from a singularity or a single big bang with earth being the most likely ontological point in time and space. Thus, the universal is most probably completely sterile of any life, since life has not seemingly evolved beyond or outside of earth. If it is out there beyond earth then it has evolved beyond earth and will share similar genetics as earth based life forms.

P.E. Alternately says that extraterrestrial life has just not been found it yet and when it is found all history books can be rewritten. If P.E. is correct, other planetary bodies will eventually demonstrate life of some sort just as earth does. The search for extraterrestrial extremophiles is a big thing now, since the finding of these little creatures in the most harsh and inhospitable terrestrial environments. P.E.'ers think that if certain bacterial forms can exist in deadly locations here, then they can exist on Mars or at least deep below the surface of the planets.

IF M.E. is correct then no matter what E.T. rock is found it will show no life at all or what life it does hide will have some connection to earthly origins.

True P.E. must demonstrate (eventually) that whatever life is found is completely dissimilar to earth life. It must prove conclusively and formally a restricted polygenesis.

The argument from the silence of evidence, evidence which is necessary to support the super abundant polygenetic origins theory rather favors greatly the truth of a monogenetic evolutionary origin. It seems that the closer the evidence is to earth the more complex life becomes . It does not get simpler.

This invalidates polygeneticism. For example (and this may be a weak argument), it should be as easy to find evidence of life in Martian rocks and in other planetary materials as earth rocks. But this is not the case. In fact, it is just the opposite. The further away from the center of life the weaker the evidence, if not a total lack of evidence. The further away from earth we get the more inhospitable space becomes to life.

All the data and the lack of evidence points to a monogenetic geo-centric origin more so than of a P.E. Origin. We have a total lack of any bio-signatures and a constant indication for a monogenetic geo-centric evolutionary origin.

So far, all evidence points to life beginning on earth. The Martian rocks and all other space materials so far are found to be sterile. There are some few chemical similarities and morphological similarities, but life seems to have "popped up" here extremely abundantly and no where else.

M.E. teaches that what life we do find can have and will most probably have a connection to earth life. There can be no other origins for life with life forms popping up and spawning elsewhere. Some NASA's scientists hope that if we prove this then we would prove P.E. To this date, it has not been done. P.E. Continues to be a hopeful potential and nothing else. P.E.'s greatest task is to first prove irrefutably the doctrine of absolute dissimilarity that there are multiple points in space where life began totally separate from earth base life forms. Obviously, this is practically impossible.

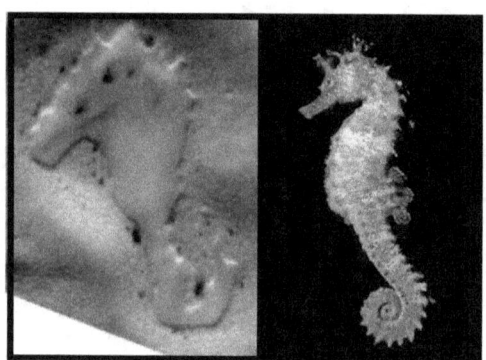

Mars Global Surveyor
MOC narrow-angle image M04-02091

STERILITY BEYOND EARTH

It seems that no matter where we send a probe to find life it returns a negative. We suggest a comical analogy, that, man will never find an odd-toed ungulate outside of earth's biosphere and therefore, he will never find an astro-horse on Mars, not even an Eohippus, a seahorse, nor any other kind of four-footed animal for that matter. But, the pro-alien life goat-ropers, for the moment, are continuing to "horse around" looking for the illusive (missing) extraterrestrial *Equidae, Equus ferus caballus*. They have failed to find one on the Moon and are now failing to find any on Mars. Why? Because, both the moon and Mars are stone dead! All lunar and Martian probes, from the Russian Luna probes through the American Lunar Ranger, Apollo and Martian landers have found no traces of life.

For example, Mariner 4 in 1965 found that Mars is too dry for any kind of life. According to NASA, the Mariner 4 mission forced most exobiologists to accept that life would not be obviously found on Mars. "The New York Times" remarked that Mars is *"probably a dead planet."* However, there was a large degree of uncertainty, because the mission had only imaged a part of the planet and had spent less than half an hour doing its work there. Thus, a new view began to emerge that life could still be present on Mars perhaps lurking in "micro-environments" such as in volcanoes, or in hot-springs somewhere. This view appears closer to our understanding of proposed Martian life today. (3a)

The Viking missions found that large floods occurred on Mars, but detected no life forms or traces of early life. Lawrence Bergreen, project scientist on the Viking Missions, said *"The mission was hailed as a great success, of course, but in terms of the search for life on Mars, it turned out to be a great disappointment. To our shock, there was nothing organic on Mars, let alone what we would call life."* Eventually, he says, they all thought, *"Well, there isn't any life on Mars. It's dead, deader than the moon."* (3b) In his testimony, he said their hopes end with them picking up their marbles, so to speak, and going home. Again, *"The Viking landers found strong oxidants in the Martian soil,"* says planetary scientist Mark Bullock of the Southwest Research Institute in Boulder, Colorado. Mr. Bullock concludes that "*Not only would these be hostile to life, but they would also consume any fossil remains of life.*" (3c) In other words, the minute any life popped up, it would be destroyed immediately!

The negative evidence continues with the Phoenix mission landing a robotic spacecraft in the polar region of Mars on May 25, 2008. Phoenix's preliminary data revealed that Martian soil contains *'perchlorate'* and thus may not be as life-friendly as thought earlier. The pH and salinity level are viewed as benign from the standpoint of biology. There was no trace of life found.

The Mars Science Laboratory mission vainly deployed the Mars rover Curiosity, a robot carrying instruments designed to look for biological activity. The Curiosity landed on Mars on Aeolis Palus in Gale Crater, near Aeolis Mons (Mount Sharp) on August 6, 2012. It has not found any biological evidence yet. It has found Martian spherules or hematite blueberries, sedimentary strata, carbonaceous material, opal, ice, lots of salts and plenty of sulfides. It has found practically everything, including kitchen sinks according to some artifact hunters, but not one microbe!

No matter how much astrobiological digging is done, the sum of all extant experiments indicate that all future experiments will only reveal negative results. Are we alone in this solar system? So far, no evidence has revealed that we are not. Not one rock turned over has proved otherwise.

Nevertheless, to maintain public support, cash flow and to hope for a P.E. proof, NASA scientists continue to probe the planets. From digging on the moon to drilling on Mars, to flying by Europa to potentially probing Uranus, and sending oceanographic submersibles to other frozen moons - the fallacious search continues for E.T. life from one failed experiment to another.

NASA GOES UNDERGROUND

It is becoming evident that it is impossible for life to exist superficially upon the surfaces of other planets other than earth. So, the groundhogs are now planning to dig below the surfaces in search for their illusive extremophile truffles. If life is not found on the surface of the planets then it must therefore be hiding underground or in sub-surface permafrosts or liquid oceans. Through Hell or high waters, sulfuric or hydrochloric ecological niches, the search continues.

Planetary scientists are now theorizing that life may exist under the cold icy surfaces of the moons of our planetary neighbors. Like teeth to be found under a pillow, the fairytale hope of finding extraterrestrial life forms is saying we will surly discover life under the ice caps of Jupiter's moons. For example, the reddish brown color of some portions of icy Europa may be caused by eruptions spewing out living (red, brown, pink?) bacteria onto the surface, which then are instantly flash frozen into what we are seeing as the reddish ting. New plans are being made to land a device to explore the underwater world thought to be below the surface of Europa, since the surface is too deadly to sustain life forms. [4]

Seems now that Mars is becoming like the moon, not too hopeful of offering any proof of extraterrestrial life, unless one digs a little deeper. What if extraterrestrial life existed in the past and has since vanished? Well, if it ain't on the surface it must be hidden below the surface, either alive or in fossil form. Amateur web sites, professional scientists, NASA and other microbial mummy hunters theorize that if there is any evidence of E.T. life it will surly be found as fossilized remains. Seems they are giving up on the false hope of finding living evidence.

As far as E.T. fossils, they say we need a way to identify possible alien remains in the search for and the analysis of fossilized extraterrestrial evidence. So, astrobiologists have spawned a new science called astropalaeobiology -- the search for extraterrestrial (microbial) fossils. This is much different from astrobiology as one only needs to dig, drill, grind, polish and then look with microscope devices for physical evidence rather than living organisms - for example, the Martian meteorites. Needless to say, the search for evidence continues with the new addition of astropalaeontology. As long as the Sun shines and it doesn't rain, the space probe equipment will never run out of energy and it can continue to search and drill holes till the cows come home or they find one on Mars!

EXAMPLES OF TERRESTRIAL ENVIRONMENTS

Of course, the theory that our planetary neighbors hide living organisms is logical, no matter the freezing cold, deadly radiation, extreme heat, poisonous and alkaline environments, and the rarity of liquid water. Contemporary reasoning seems to go like this: Microbes have flourished on earth for more than 3.5 billion years and in the most deadly environments too. They thrive in the most inhospitable earth conditions such as beneath ice sheets in Antarctica. Therefore, since Mars is an extremely cold planet similar to Antarctica, having large ice deposits, it is possible that similar microbial organisms might be found there. (4a)

EARTH ANALOGOUS TO OTHER PLANETS

Palaeobotanists find micro-fossil evidence in early Devonian sedimentary (chert) deposits of the prehistoric hotbed ecosystems of Rhynie and Windyfield Scottland. They have found living extremeophiles in geochemically active extreme conditions that are detrimental to most life on Earth. [5] For example, they find polyextremophile thermophilic and barophilic organisms living inside hot rocks deep under Earth's surface. [6] There are also the pH tolerant radioresistant xerophiles, psychrophiles and oligotrophes living at the summit of a mountain in the Atacama Desert. [7] Also found are microbes living in liquid asphalt in Pitch Lake at La Brea in southwest Trinidad. [8] Other "things" are found living in ice 3,700 metres (12,100 ft) deep at Lake Vostok in Antarctica [9], as well as weird little Frankensteinish things in boiling water, sulfuric acid and in the water core of nuclear reactors, not to mention other stuff living in salt crystals, toxic waste and a whole range of other extreme habitats that were previously thought to be deadly to life.[10] So, why not in the extreme conditions on Mars and other planets? Though all of these habitats usually rest outside the "habiltible zone" such as where earth is, they might, under certain conditions offer habitable niches.

Is it possible for life to exist on some of these bodies? It is important to remember that the habitable zone, by definition, is the region where a planet could potentially have surface temperatures that would support liquid water, but a more accurate name for this zone might be the "zone of liquid water," as scientists now believe that habitability can occur outside this zone in certain conditions.

For example, NASA believes astro-varmints may survive in subsurface groundwater, which may still exist on Mars today as well as in subsurface oceans like the one believed to exist on Jupiter's moon Europa. Recent images taken by NASA's Cassini Spacecraft show what may be liquid water reservoirs that erupt in Yellowstone-like geysers on Saturn's moon Enceladus. Some believe that, even if the waters are found to be super acidic, maybe the little varmints their evolved into sulfuric and other acidic resistant forms.

POSSIBLE ASTROBIOLOGICAL ENVIRONMENTS

The above terrestrial examples have provided data for extrapolations to the likelihood of microorganisms surviving frozen in extraterrestrial habitats. Therefore, in search of unearthly life forms, living microorganisms are the most likely candidates for a biota of an extraterrestrial habitat. Thus, by analogy with earthly extremophiles, potential niches or biological oases have been postulated for Mars [11] and other planetary systems. They may potentially live on the planets Venus and Mars, along with several natural satellites orbiting Jupiter and Saturn, and even comets are suspected to possess niche environments in which life might exist. A subsurface marine environment on Jupiter's moon Europa might be the most suitable habitat in the Solar System, outside Earth, for multicellular organisms. [12]

It follows that Mars and other water and ice worlds may harbor endolithic communities or biotic oases where planetary parasites, astro-amoebas and exo-skeletal micro-cryocrabs may exist, such as in deep subsurface permafrost, subsurface oceans and geothermal vents. Even the subsurface water ocean of Europa may harbor life, especially at hypothesized hydrothermal vents at the ocean floor. Microbes (extremophiles) could also exist in the stable cloud layers 50 km (31 mi) above the surface of Venus [13]. The desparate attempt to find alien life goes so far as peeking into cosmic dust water droplets, potential asteroidial ecological niches and other far-off, out of the way, unlikely places.

Hell, if earth based bacteria (Streptococcus mitis) can survive in the lunar Surveyor 3 camera for 2 years, until Apollo-12 Astronauts found them and brought them back home (13a), then they can live anywhere! Right? Sorry to say, the answer is, "No." There is a big difference in survival times between 2 years and billions of years when considering radiation.

The Wikipedia article "Life on Mars" tells us that life on Mars, if it ever existed, would have only lasted at the most 1/2 million years under present conditions: "Currently, ionizing radiation on Mars is typically two orders of magnitude (or 100 times) higher than on Earth. (13b) Even the hardiest cells known could not possibly survive the cosmic radiation near the surface of Mars for that long. (13c) After mapping cosmic radiation levels at various depths on Mars, researchers have concluded that any life within the first several meters of the planet's surface would be killed by lethal doses of cosmic radiation. The team calculated that the cumulative damage to DNA and RNA by cosmic radiation would limit retrieving viable dormant cells on Mars to depths greater than 7.5 metres below the planet's surface." And, "Even the most radiation-tolerant Earthly bacteria would survive in dormant spore state only 18,000 years at the surface; at 2 meters —the greatest depth at which the ExoMars rover will be capable of reaching— survival time would be 90,000 to half million years, depending on the type of rock." (13d)

In other words, if life ever began on Mars, it died 1/2 million years (at the latest) after the Martian Hesperian Period (named after Hesperia Planum): between 3.0 to 3.7 billion years ago. This fits exactly to the time period that life supposedly originated on earth 3.6 billion years ago, with the appearance of simple cells (prokaryotes). (13e) The oldest known fossilized prokaryotes were (are believed to have been) laid down approximately 3.5 billion years ago, only about 1 billion years after the formation of the Earth's crust 4.6 billions years ago. The origin of these simple cells or prokaryotes are said to have evolved out of (we cannot explain) self-replicating molecules, which (somehow?) formed proto-cells or "primitive" cells - assemblages of essential components or vesicles (somehow?) encapsulated and (someway?) cooperating, and (coincidentally) interacting within an enclosed membrane (that we do not know how developed). How long this took no ones as yet knows. Did Mars have time to produce its own versions of prokaryotes? If so, it had only one billion years to do it and no proof has been found yet to affirm this.

Some scientists say, that Mars could have had and might still have ecological niches where lichens live and thrive. The German Aerospace Center's Institute of Planetary Research in Berlin ran an experiment proving lichens can live in a simulated Martian environment surviving temperatures as low as -51 degrees C and enduring a radiation bombardment during a 34-day experiment. But the lichen, a symbiotic mass of fungi and algae, also proved it could adapt physiologically to living a normal life in such harsh Martian conditions. Of course, this is only true as long as the lichen lived under "protected" conditions shielded from much of the radiation within "micro-niches" such as cracks in the Martian soil or rocks. (13f)

Unfortunately, lichens probably do not exist on Mars. They are a much later development in the supposed evolutionary chain and would not have developed after the beginning of the extreme harsh conditions of the Hesperian Period and the extinction of the Martian prokaryotes, from which they would have derived. No prokaryotes means no lichens!

The above logic of the space microbe hunters runs as follows: Earth has extremeophiles and simple cells that live in extreme and even deadly envirnments. Mars has extreme and even deadly environments. Therefore, Mars might very well have extremeophiles!

MARS HAS COCKROACHES

If we allow ourselves to believe this, then Cockroaches live on Mars. It was said cockroaches were found to have survived the bombing of Hiroshima, Japan. This was questioned later and because no one was really sure if the Japanese cockroaches really went through the bombing, a test was performed. An experiment was performed by a group of scientists (13g) where cockroaches were subjected to high doses of radiation to test whether it was true that cockroaches could survive an atomic blast.

Now, since the bomb on Hiroshima emitted radioactive gamma rays at a strength of around 10,000 rads, the roaches were radiated for a 30 day period. The results showed that half the roaches exposed to 1,000 rads were still kicking, and a remarkable 10 percent of the 10,000 rad group was alive. The results confirmed that cockroaches can survive a nuclear explosion.

Well, what we have here is proof that cockroaches might very well be living on Mars! According to the above logic, if cockroaches can survive an atomic bomb radiation blast on earth, they can survive the radiation bombarding the surface of Mars. We have shown that Earth has radioactive resistant cockroaches. Mars has radiation too. Therefore, Mars could have radioactive resistant cockroaches. Seems reasonable doesn't it?

What about the radiation-resistant cocci "Conan the Bacterium" belonging to the genus Deinococcus radiodurans from sewage sludges and animal feeds? Six strains of radiation-resistant gram-positive cocci were isolated from sewage sludges and animal feeds in Japan after gamma-irradiation of more than 1.0 Mrad. All six strains were able to growon nutrient agar slants, and somestrains were also able to growon glutamate agar slants. [13h] How about Bdelloid rotifers - two species shrugged off as much as 1,000 Grays and were still active two weeks after exposure.[13i]

A Brazilian study in a hill in the state of Minas Gerais which has high natural radiation levels from uranium deposits, has also shown many radioresistant insects, worms and plants. [13j] Turtles can swallow 15Gy. Goldfish can swim in 20Gy and Escherichia coli is ablivious to 60Gy. German cockroaches can twitch threw about 64Gy, while Shellfish are hardened against 200Gy. The Fruit fly flies threw some200+Gy, but the Amoeba has it beat at 1000Gy. The blue ribbon winners of higher Gray levels are Braconidae up to 1800Gy and Milnesium tardigradum at up to lethal doses of 5000Gy. The top world champs of chomping on high Rads are Deinococcus radiodurans (previously talked about) and the global champ, Thermococcus gammatolerans (T.G.). Conan the Bacterium was not the most radiation resistant organism that we know of anymore. T.G. was discovered in 2003 near hydrothermal vents in the Guaymas Basin and it can withstand 30 thousand Gy! [13k]

ASSESSING THE PROBABILITY OF THE EXISTENCE OF COCKROACHES IN OUR GALAXY

$N = R^* \times Fp \times Ne \times Fl \times Fi \times Fc \times L$

Using the above "Drake" formula, where:

N = the number of cockroaches in our galaxy with which we might hope to be able to find.

R* = the average rate of star formation in our galaxy = 1.025 Msun/yr. [Total star formation rate = 0.68 to 1.45 Msun/yr. = Averaged "mean" = .60 + 1.45 = 2.05 div by 2 = 1.025 Msun/yr. See, http://arxiv.org/abs/1001.3672]

Fp = the fraction of those sunlike stars that have planets. [= 20% See, http://www.extremetech.com/extreme/170404-kepler-20-of-sun-like-stars-have-habitable-planets-alien-life-drake-equation-finally-has-a-leg-to-stand-on]

Ne = is the average number of planets that can potentially support life per star that has planets. [= 60 billion. See, http://www.space.com/21800-alien-planets-60-billion-habitable-exoplanets.html]

Fl = the fraction of the above that actually go on to develop simple life forms at some point. [= pick random estimate = .01% of 60 billion = 6,000,000.ooo]

Fi = the fraction of the above that actually go on to develop cockroaches. [= random estimate = .01% of 60 mill = 600]

Fc = the fraction of cockroaches that develop a large enough population that releases detectable biosigns (gases, fumes, stinches) of their existence into space. [= random estimate = again, .01% of 600 = .06]

L = the length of time such developed cockroach populations release detectable signals into space. [= 3.6 billion years]

A population of organisms grows at a rate that depends on how many organisms are currently in the population. For instance, if there 100 cockroaches:

Well, suppose each mother cockroach can have 10 baby cockroaches per month. For simplicity, let's also say that once a cockroach has babies, it then dies. If you started with 100 cockroaches, after a month you would have 1,000 cockroaches (each of the 100 would have 10 babies, and 100 * 10 = 1,000). If you started with 1,000 cockroaches, after a month you would have 10,000 (each of the 1,000 had 10 babies, and 1,000 * 10 = 10,000).

Starting with 1,000 cockroaches, you end up with 10 times as big a population at the end of the month as you would be starting from just 100 cockroaches. But despite the different population sizes, both populations of cockroaches grew at the same rate. That rate was 10 new cockroaches for each old cockroach during the month. This is known as the "intrinsic growth rate" of the cockroaches, and this kind of unlimited growth is called "exponential growth."

In equations, exponential growth looks like this:

Population (time + 1) = growth_rate * Population (time)

This just says that the number of individuals one time unit in the future (one month, in our example) will be equal to the number of individuals now times the growth rate of the population. If the adults don't die before the kids grow up, then the formula becomes slightly more complicated because we have to add today's adults into the next time's population:

Population (time + 1)
= Population(time) + growth_rate * Population(time)

This is the same as the first formula, but here the population at the next time point is made up of both the babies
born and todays adults.

Exponential growth can make a population grow very quickly. Let's start with the 100 cockroaches and grow them for twelve months. We will see that in one year, there will be 1e+26 roaches.

Number of Months	Starting population	Ending population
1	100	1,000
2	1,000	10,000
3	10,000	100,000
4	100,000	1,000,000
5	1,000,000	10,000,000
6	10,000,000	100,000,000
7	100,000,000	100,000,000,000(b = billion)
8	100(b)	100 trillion
9	100t	100[15] or 1e+17
10	100[15]	100[18] 0r 1e+20
11	100[18]	100[21] or 1e+23
12	100[21]	100[24] or 1e+26

Multiply the further over a continued 3.6 billion years, or the estimated time it took "mankind" to evolve and this will give us an estimated total population of cockroaches.

$$N = R^* \times Fp \times Ne \times Fl \times Fi \times Fc \times L$$

CALCULATING THE ABOVE:

R*	1.025 Msun/yr
x Fp	x 20%
x Ne	x 60,000,000,000
x Fl	x 6,000,000,000
x Fi	x 600
x Fc	x .06
x L	x 3.6 billion years
	TOTAL = 9.56448e+21 = N

JUMPING TO JUPITER: THE SEARCH CONTINUES

There are many extreme environments that might host life forms as well as many possible extraterrestrial environments such as the moons of Jupiter and Saturn, Neptune, the asteroid belt and other planetoids.

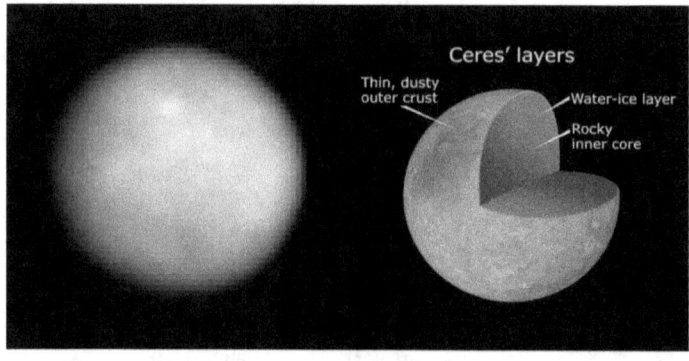

CERES

Ceres is the largest asteroid and the only dwarf planet in the inner Solar System (asteroid belt). It was discovered in 1801 by Giuseppe Piazzi. It is named after Cerēs, the Roman goddess of growing plants, the harvest and motherly love. Ceres is recently confirmed to have water vapor in its atmosphere. Frost on the surface has also have been detected.[14] The presence of water and the temperatures on Ceres make it possible that life may exist there.[15] The Dawn space probe is scheduled to enter orbit around Ceres in spring 2015.[16]

Some speculate that Ceres may offer extreme habitats for life. Others suggests it's too damn cold. But, Ceres is gushing water vapor from its unusual ice-covered surface. This raises the question of whether it might be hospitable to life. The spewing water vapor caused by the Sun warming parts of the icy surface provides some proof that Ceres may have some kind of atmosphere. There may be liquid water under the frozen surface of Ceres and that may be the origin of the vapor shooting out of the geysers or icy volcanoes.

Scientists think Ceres holds rock in its interior and is wrapped in a mantle of ice that, if melted, would amount to more fresh water than is contained on Earth. Ceres is one of the few places in the solar system aside from Earth where water has been located. A big question is what this water vapor means regarding the possibility of life.

There's a lot more ingredients than just water that's required for life. And whether Ceres has those other ingredients, it's too early to say.(16a) So far no signs of life and no biosignatures have been found.

MORE MOON MICROBES

The heated subsurface oceans of water that are speculated to exist deep under the crusts of the four outer moons of Jupiter -- Ganymede, Callisto, Europa and Io are also likely locations for life forms, at least according to the polygenists.

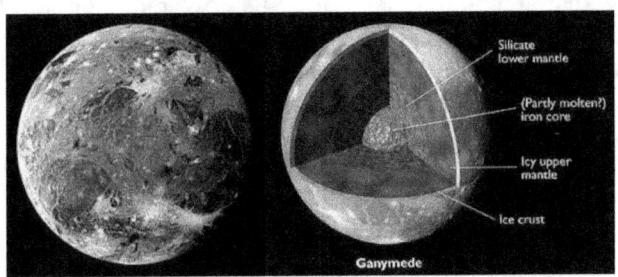
Ganymede

GANYMEDE

Ganymede, one of the moons of Jupiter, is believed to be a likely candidate for life. It houses a saltwater ocean sandwiched between layers of ice some 200 km below the surface.[17] This is a very tempting environment for life forms to exist. Ganymede is, like Europa, a large, ice-covered moon. It too has a subsurface ocean, which could potentially host life. Whether it exists in pockets or as a continuous band around the moon, are questions to answer. Ganymede is more complex than all other moons. It has a magnetic field and probably a liquid-iron core powerful enough to generate an aurora, like Earth's. It also has a tenuous oxygen atmosphere formed by the breakdown of water ice on the surface. This moon appeals to geologists, astrobiologists, magnetophysicists and atmospheric scientists because it is clearly a very rich environment. [17a] No life or signs of life have been found to date.

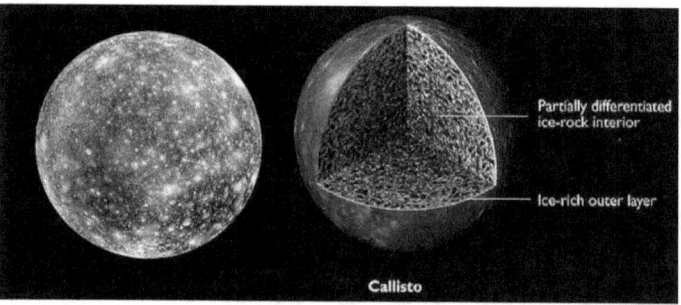

Callisto

CALLISTO

Callisto, another moon of Jupiter, also offers some hope of finding life forms. Its composition is about half water ice and half rocky material. it consists of water, magnesium, iron-bearing hydrated silicates, carbon dioxide, sulfur dioxide, possibly ammonia and other organic compounds. The presence of an inner ocean leaves open the possibility that it could sustain life. Its low radiation levels present Callisto as being considered a suitable place for future exploration. [18] Again, so far, no life or bio-signs have been found.

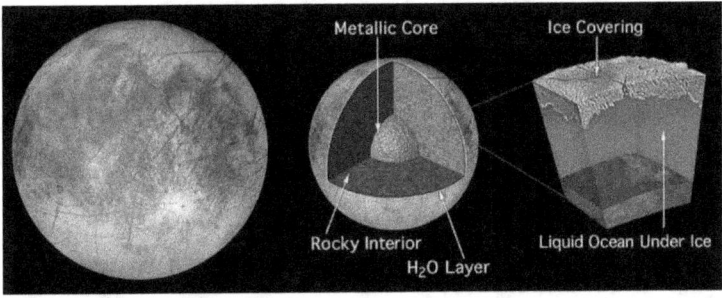

EUROPA

Europa, one of the more hopeful moons of Jupiter, on the other hand is mostly frozen water. It absorbs infrared radiation differently than normal ice and thus displays a funny red color. This is not normal and something is causing this oddity. Researchers think this is because something is binding the water molecules together. For example, salts of magnesium sulphate would make the molecules vibrate at different frequencies. Another explanation would be bacteria. This red tinge of Europa could be caused by frozen bits of bacteria. This could explain Europa's mysterious infrared signal.

Bacteria cannot survive on Europa's surface, but they could live in the liquid water inside Europa's icy crust. Bacteria could be blasted out to the surface in some kind of eruption and flash frozen, thus causing the reddish color. [18a] Nice moon, but again, scientists have not yet found any living microbes and may never find any.

Europa could be flat dead, say some scientists! Chemicals called oxidants found on the surface of Europa might jeopardize chances of life evolving there. The level of acidity in its ocean cannot be friendly to life. Acidity messes with delicate membrane development and the building of large-scale organic polymers.

These destructive oxidants are rare in the solar system because of the abundance of reductant chemicals such as hydrogen and carbon that react with these oxidants to form oxides such as water and carbon dioxide. This is how planet earth works. Europa happens to be different. It does not have these reductant elements and so is super rich in strong oxidants such as oxygen and hydrogen peroxide, which are created by the irradiation of its icy crust by high-energy particles from Jupiter. Also, the oxidants on Europa's surface are likely carried downward in potentially substantial quantities by the same churning that causes water to rise from below. This possibly causes a reaction with the sulfides and other compounds in its interior ocean generating sulfuric and other acids.

If this has occurred for just about half of Europa's lifetime, not only would such a process rob the ocean of life-supporting oxidants, but it would become relatively corrosive, with about the same pH level (2.6) of a soft drink or what we call a soda-pop. [19] Instead of oceans full of water, just imagine swimming for long periods of time in an ocean of Pepsi Cola!

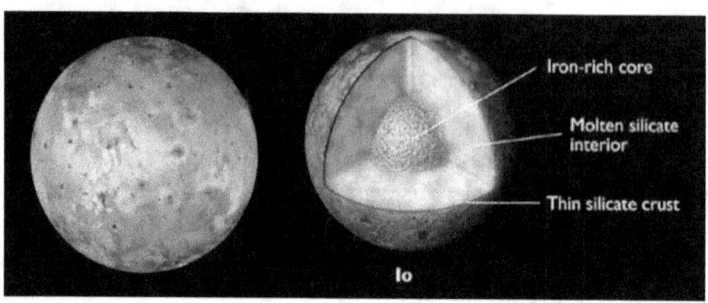

IO

Io is the innermost of Jupiter's large satellites and the most volcanically active body in the solar system. It is considered a possible candidate as a hot spot for extreme extraterrestrial (theoretical) sulfur compound based life forms. Conditions on Io might have made it a friendlier habitat in the distant past for carbon based life forms as we know them. But, life on the surface as it is now is biologically impossible.

It is categorically dead right away because of it's extreme hydrogen sulfide environment. No organic molecules have been detected on Io's surface. And no wonder, any organic compounds that once existed on the surface or that may today still emanate from the subsurface would be quickly destroyed by Jupiter's radiation. Jupiter's radiation stripped all water from Io's surface leaving it deadly to life forms.

This does not mean they do not exist underground. Whatever developing life there might have been could have easily retreated underground, where water might still be abundant, and geothermal activity and sulfur compounds could provide microbes with sufficient energy to survive.

If probes could dig down further into the rocks of Io, they might possibly find life-forms. Microbes are common deep under ground and in lava tubes on Earth. If lava tubes exist on Io, they could also serve as an especially favorable extraterrestrial environment for life by protecting organisms from Jupiter's radiation and providing thermal insulation, trapping moisture and providing nutrients such as sulfurous compounds. [19a] Life is absent from the surface. Future missions will tell for sure if Io has any life.

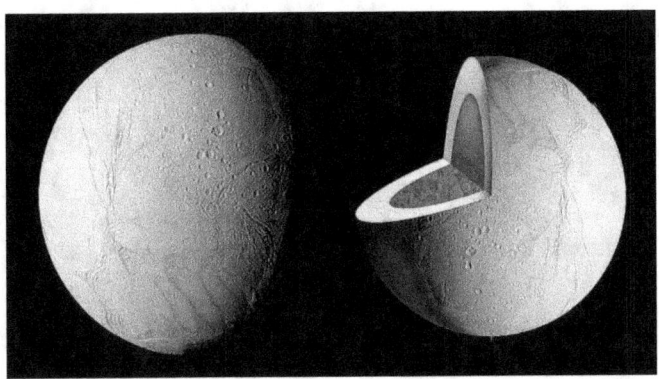

ENCELADUS

Enceladus is the sixth-largest of the moons of Saturn. It seems to have liquid water under its icy surface and water-containing plumes or geysers at the south pole shooting large jets of water vapor, other volatiles and some solid particles into space. This is very similar to those of Jupiter's moons Europa and Io and Neptune's moon Triton with its nitrogen "geysers."

The outgassing of these plumes obviously originates from a body of subsurface liquid water, and the unique chemistry found in the outgassing plumes suggests the possibility that Enceladus may be important as the most habitable spot beyond Earth in the Solar System for life as we know it.[20] Until a lander tests the surface and interior waters, we are left to speculation as to the presence of life.

TITAN

Titan is the sixth moon of Saturn and the second largest moon in the solar system, after Jupiter's moon Ganymede. Titan has a diameter 50% larger than Earth's natural satellite, the Moon, and is 80% more massive. It is the only natural satellite known to have a dense atmosphere,[21] and the only object other than Earth with a stable body of surface liquid.[22] It is primarily composed of water ice and rocky material and has a dense, opaque atmosphere preventing any view of its surface. The Cassini–Huygens mission in 2004, discovered liquid hydrocarbon lakes in Titan's polar regions.

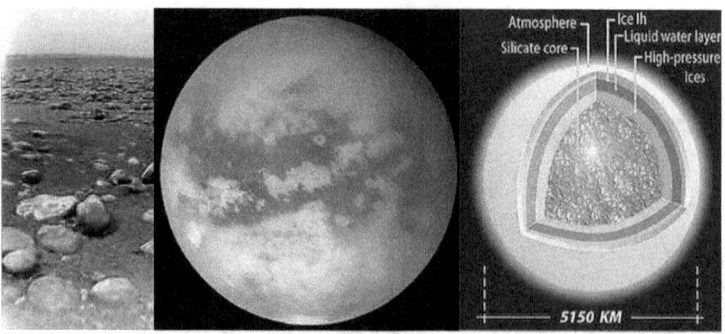
Image of Titan's surface. Huygens probe, January 14, 2005

The geologically young surface is generally smooth, with few impact craters, mountains and several possible cryovolcanoes.[23] The atmosphere of Titan is largely composed of nitrogen, methane, ethane clouds and nitrogen-rich organic smog. The climate creates surface features similar to those of Earth, such as dunes, rivers, lakes, seas (made of liquid methane and ethane), and deltas.

Titan is thought to be a prebiotic environment rich in complex organic chemistry[24] with a possible subsurface liquid ocean serving as a potential biotic environment.[25] The Cassini–Huygens mission showed an environment on Titan that is similar, in some ways, to ones theorized for the primordial Earth.[26]

One model suggests an ammonia–water solution deep beneath a water-ice crust with conditions that could support life. Detection of microbial life on Titan would depend on its biogenic effects. That the atmospheric methane and nitrogen might be of biological origin has been examined.[27] Life could exist in the lakes of liquid methane on Titan, just as organisms on Earth live in water[28] if one speculates that life on Titan uses instead a liquid hydrocarbon, such as methane or ethane, rather than water as a liquid solvent.[29] Such creatures would inhale H_2 in place of O_2, metabolize it with acetylene instead of glucose, and exhale methane instead of carbon dioxide.

There are formidable obstacles to life on Titan. At a vast distance from the Sun, Titan is frigid, its atmosphere lacks CO_2 and water exists only in solid form. Because of these difficulties,Titan is less likely to be a habitat for life. NASA notes in its news article on the June 2010 findings: "*To date, methane-based life forms are only hypothetical. Scientists have not yet detected this form of life anywhere*".[30]

NEWS AND ENTERTAINMENT MISINFORMATION

Needless to say, these planetary systems make for exciting speculations about life on other worlds. This frenzy of false hope and misinformation of unearthly life forms is demonstrated in such sci fi movies as Operation Ganymede (1977), 2010, The Year We Make Contact, Buddy Holly is Alive And Well on Ganymede, Night Caller From Outer Space (Horror) and Blood Moon (2008). The movie Outland is also set around Io, another moon of Jupiter, where mining operations are being done, while "The Europa Report" depicts astronauts finding life forms under the ice sheet and barely escaping being a Gany-meal.

This theme of life on other planets is continued in new TV series and even dates back as early as black and white silent films: Escape from Jupiter (1994); Le voyage sur Jupiter (1909 short). The list of Hollywood exo-planetary prokaryotic life forms goes on and beyond the moons of Jupiter. The classic sci fi movie Green Slime depicts a large astroid (from where, we do not know) headed toward earth oozing with a photosynthetic green slimy microbial substance that enjoys electrocuting people.

Apparently, from the above short lists of both government and entertainment industries, the idea of E.T. life is a given truth based upon an undying belief that life exists beyond earth and man must find it - and at all costs, too! There is no need to emphasize this fact that, as Rovers rampage across the Martian surface, many children suffer malnutrition even in our own country. But, never mind the gaunt faces of impoverished children hiding behind the black-budget curtains of the alien presence theory, the show must go on! The search for the Holy Green-Slime Grail must be sought!

In volume-1 "Is Anyone Else on the Moon" we looked at the Moon and Mars as potential places for life to dwell. As anticipated, we found the moon to be dead, Mars to be sterile and both lacking alien artifacts, even though the yarn-spinners keep producing blurry and fuzzy objects depicting alien junk, motor parts, screws, square-things and even faces and big-foot looking creatures staring at the Rover's camera. From the above synopsis of our other planets and their moons we see that life is still yet to be found. Most every planetary system we have studied has demonstrated a lack of life.

The problem is that none of these extraterrestrial planets or moons has the proper "soil" like environment as that of earth. I propose this as the problem. If you are going to have life, you first have to have the proper environment or "soil." None of these planets or moons have the proper "soil" or environment to produce or maintain life. It might be predicted that in the future, just as it has been in the past, no life will be found. No soil, no seed growth. If there ever were microbes on these cosmic orbs, just as soon as they evolved and began to thrive, they all would have died immediately,

With such a solar system wide lack of microbial life, bio-signatures and even fossil remains, one should get the message that exobiological life forms do not exist. This should tell us something about the potential reality of alien cultures, their developing on other habitable exo-planets and their coming from other planets to visit humans and, best of all, traveling trillions of miles only to kidnap humans and probe their bums!

But these points do not stop people from searching for advanced space perverts and microscopic exobiological extremophiles. After Leonard hatched his "Somebody Else Is on the Moon" book, others followed with more hyped up pseudoscientific pro-alien studies, unthwarted by the conventional scientific warnings that there are probably no space beings.

In this volume we will critique Mr Fred Steckling's bizarre discourse on alien bases on the moon, a few other E.T. trash and garbage hunters and then end with the laughable myth-making pareidolean apophenia (seeing things that aren't there) theories of Richard Hoagland.

FOOTNOTES

[1] http://www.space.com/video

[2] http://www.space.com/24834-strange-mars-meteorite-life-evidence-debate.html

[3] http://news.discovery.com/space/mystery-rock-appears-in-front-of-mars-rover-140117.htm.
And, http://www.space.com/24356-mars-rock-mystery-opportunity-rover-photos.html

[3a] Mariner 4: "First Spacecraft to Mars." by Elizabeth Howell, SPACE.com Contributor. See, www.space.com/18787-mariner-4.html

[3b] Bergreen, Lawrence. (2000) "Voyage to Mars: NASA's Search for Life Beyond Earth." Riverhead Books, N.Y. P.188-189.

[3c] :Phoenix discovery may be bad for Mars life." Maggie McKee, space editor. NewScientist.com.
http://www.newscientist.com/blog/space/2008/08/phoenix-discovery-may-be-bad-for-mars.html

[4] NewScientist.com under search "Europe."

[4a] "Habitability and Biology" by The Phoenix Mars Mission. See, http://phoenix.lpl.arizona.edu/mars143.php

[5] Rampelotto, P. H. (2010). Resistance of microorganisms to extreme environmental conditions and its contribution to Astrobiology. Sustainability, 2, 1602-1623.
Rothschild, L.J.; Mancinelli, R.L. Life in extreme environments. Nature 2001, 409, 1092-1101

[6] Oskin, Becky (14 March 2013). "Intraterrestrials: Life Thrives in Ocean Floor". LiveScience. Retrieved 17 March 2013. Also, Nanjundiah, V. (2000). "The smallest form of life yet?". Journal of Biosciences 25 (1): 9–10.

[7] Atacama Desert, a plateau in South America, covering a 1,000-kilometre (600 mi) strip of land on the Pacific coast, west of the Andes mountains.

[8] Astrobiology, Vol. 11, p. 241-258)

[9] "Detection, recovery, isolation, and characterization of bacteria in glacial ice and Lake Vostok accretion ice". Ohio State University. 2002.

[10] Carey, Bjorn (7 February 2005). "Wild Things: The Most Extreme Creatures". Live Science.
Cavicchioli, R. (Fall 2002). "Extremophiles and the search for extraterrestrial life". Astrobiology 2 (3): 281–92.
The BIOPAN experiment MARSTOX II of the FOTON M-3 mission July 2008.
Surviving the Final Frontier. 25 November 2002.

[11] Horneck, Gerda. "The microbial world and the case for Mars"- Planetary and Space Science 48 (2000) 1053-1063

[12] Wikipedia, under search "Extraterrestrial Life."

[13] Venusian Cloud Colonies :: Astrobiology Magazine Geoffrey A. Landis Astrobiology: The Case for Venus. Cockell, C. S. (December 1999). "Life on Venus". Planetary and Space Science 47 (12): 1487–1501.

[13a] "Earth microbes on the Moon". Science.nasa.gov.

[A common type of bacterium, *Streptococcus mitis*, accidentally contaminated the Surveyor's camera prior to launch, and the bacteria survived dormant in the harsh lunar environment for two and one-half years, supposedly then to be detected when Apollo 12 brought the Surveyor's camera back to the Earth. This claim has been cited by some as providing credence to the idea of interplanetary panspermia, but more importantly, it led NASA to adopt strict abiotic procedures for space probes to prevent contamination of the planet Mars and other astronomical bodies that are suspected of having conditions possibly suitable for life. See Wikipedia, under "Surveyor 3"]

[13b] Keating, A.; Goncalves, P. (November 2012). "The impact of Mars geological evolution in high energy ionizing radiation environment through time". Planetary and Space Science – Eslevier 72 (1): 70–77.

[13c] Than, Ker (January 29, 2007). "Study: Surface of Mars Devoid of Life". Space.com. "After mapping cosmic radiation levels at various depths on Mars, researchers have concluded that any life within the first several yards of the planet's surface would be killed by lethal doses of cosmic radiation." And,

Dartnell, Lewis R.; Storrie-Storrie-Lombardi, Michael C.; Muller, Jan-Peter; Griffiths, Andrew. D.; Coates, Andrew J.; Ward, John M. (2011). "Implications of Cosmic Radiation on the Martian Surface for Microbial Survival and Detection of Fluorescent Biosignatures"

[13d] Dartnell, L. R.; Desorgher, L.; Ward, J. M.; Coates, A. J. (2007). "*Modelling the surface and subsurface Martian radiation environment: Implications for astrobiology*". Geophysical Research Letters 34 (2): L02207. Bibcode:2007GeoRL..34.2207D. doi:10.1029/2006GL027494. "*Bacteria or spores held dormant by freezing conditions cannot metabolise and become inactivated by accumulating radiation damage. We find that at 2 m depth, the reach of the ExoMars drill, a population of radioresistant cells would need to have reanimated within the last 450,000 years to still be viable. Recovery of viable cells cryopreserved within the putative Cerberus pack-ice requires a drill depth of at least 7.5 m.*" And,

Lovet, Richard A. (February 2, 2007). "Mars Life May Be Too Deep to Find, Experts Conclude". National Geographic News. "*That's because any bacteria that may once have lived on the surface have long since been exterminated by cosmic radiation sleeting through the thin Martian atmosphere.*"

[13e] http://en.wikipedia.org/wiki/Timeline_of_evolutionary_history_of_life

[13f] Lichens on Mars, Astrobiology Magazine:

http://www.astrobio.net/exclusive/5932/lichen-on-mars

[13g] "Cockroaches Survive Nuclear Explosion" http://www.discovery.com/tv-shows/mythbusters/mythbusters-database/cockroaches-survive-nuclear-explosion.htm

(13h) http://ci.nii.ac.jp/naid/130000027414/.

[13i] http://blogs.discovermagazine.com/notrocketscience/2008/03/24/bdelloid-rotifers-the-worlds-most-radiation-resistant-animals/#.U0naWaKwUY0

[13j] http://piecubed.co.uk/conan-bacterium-radiation/

[13k] http://en.wikipedia.org/wiki/Radioresistance

Cordeiro, AR; Marques, EK; Veiga-Neto, AJ (1973). "Radioresistance of a natural population of Drosophila willistoni

living in a radioactive environment.". Mutation research 19 (3): 325–9

Moustacchi, E (1965). "Induction by physical and chemical agents of mutations for radioresistance in

Saccharomyces cerevisiae". Mutation research 2 (5): 403–12

[14].http://www.panspermia-theory.com/extremophiles/extremophiles-and-panspermia

[15] Küppers, M.; O'Rourke, L.; Bockelée-Morvan, D.; Zakharov, V.; Lee, S.; Von Allmen, P.; Carry, B.; Teyssier, D.; Marston, A.; Müller, T.; Crovisier, J.; Barucci, M. A.; Moreno, R. (2014-01-23).

"Localized sources of water vapour on the dwarf planet (1) Ceres". Nature 505 (7484): 525–527.

Campins, H.; Comfort, C. M. (2014-01-23). "Solar system: Evaporating asteroid". Nature 505 (7484): 487–488.

A'Hearn, Michael F.; Feldman, Paul D. (1992). "Water vaporization on Ceres". Icarus 98 (1): 54–60.

O'Neill, Ian (5 March 2009). "Life on Ceres: Could the Dwarf Planet be the Root of Panspermia". Universe Today. Retrieved 30 January 2012.

Catling, David C. (2013). Astrobiology: A Very Short Introduction. Oxford: Oxford University Press. p. 99.

Is there life on Ceres? Dwarf planet spews water vapor into space. (22 January 2014).

[16] "Dawn has departed the giant asteroid Vesta". NASA/JPL. 2012-09-05.

[16a] "Water vapor plumes raise question about life on dwarf planet Ceres." By Will Dunham, WASHINGTON Wed Jan 22, 2014.

Ref: http://www.reuters.com/article/2014/01/23/us-space-ceres-idUSBREA0M02J20140123

[17] Solar System's largest moon likely has a hidden ocean". Jet Propulsion Laboratory. NASA. 2000-12-16.

[17a] NewScientist.com "Why Jupiter's moon Ganymede is an exciting destination" 15:30 3 May 2012. Justin Mullins, contributor.

[18] Space.com "Callisto: Facts about Jupiter's Dead Moon" Kim Zimmermann

[18a] NewScientist.com "Bacterial explanation for Europa's rosy glow" 10:34 11 December 2001 by Nicola Jones

[19] "Jupiter Moon's Ocean May Be Too Acidic for Life."

Charles Q. Choi, Astrobiology Magazine Contributor. March 01, 2012

[19a] "Is there Life on Europa, Io or Ganymede?" New Mission Set to Jupiter's Moons.
http://www.dailygalaxy.com/my_weblog/2012/05/life-on-europa-io-or-ganymede-new-mission-to-jupiters-moons-.html

[20] Lovett, Richard A. (31 May 2011). "Enceladus named sweetest spot for alien life". Nature (Nature). doi:10.1038/news.2011.337.

[21] "News Features: The Story of Saturn". Cassini–Huygens Mission to Saturn & Titan. NASA & JPL.

[22] Stofan, E. R.; Elachi, C.; Lunine, J. I.; Lorenz, R. D.; Stiles, B.; Mitchell, K. L.; Ostro, S.; Soderblom, L. et al. (2007). "The lakes of Titan". Nature 445 (1): 61–64.

[23] "NASA Titan - Surface". NASA. Retrieved 2013-02-14. And, G Mitri (2007). "Hydrocarbon lakes on Titan".

[24] Staff (April 3, 2013). "NASA team investigates complex chemistry at Titan". Phys.Org.

[25] Titan is thought by some scientists to be a possible host for microbial extraterrestrial

life. Grasset, O.; Sotin, C.; Deschamps, F. (2000). "On the internal structure and dynamic of Titan". Planetary and Space Science 48 (7–8): 617–636. And,

Fortes, A. D. (2000). "Exobiological implications of a possible ammonia-water ocean inside Titan". Icarus 146 (2): 444–452. Also, McKay, Chris (2010). "Have We Discovered Evidence For Life On Titan". SpaceDaily. Retrieved 2010-06-10. Space.com. March 23, 2010.

[26] Raulin, F. (2005). "Exo-astrobiological aspects of Europa and Titan: From observations to speculations". Space Science Review 116 (1–2): 471–487.

[27] Fortes, A. D. (2000). "Exobiological implications of a possible ammonia-water ocean inside Titan". Icarus 146 (2): 444–452.

[28] McKay, C. P.; Smith, H. D. (2005). "Possibilities for methanogenic life in liquid methane on the surface of Titan". Icarus 178 (1): 274–276.

[29] Committee on the Limits of Organic Life in Planetary Systems, Committee on the Origins and Evolution of Life, National Research Council; The Limits of Organic Life in Planetary Systems; The National Academies Press, 2007; page 74.

[30] "What is Consuming Hydrogen and Acetylene on Titan?". NASA/JPL. 2010.

SECTION-3
Did We Discover Alien Bases on the Moon?
FRED STECKLING Sticks It to Us!
A Refutation of Fred Steckling's
"We Discovered Alien Bases on the Moon"

FRED STECKLING
On Nationwide Television (1977)

The second most important study in the chain of alien artifact hunting lunacy, demonstrating alien moon monuments, is Fred Steckling's *"We Discovered Alien Bases on the Moon."*

In this small tome of extra-terrestrial Tommy-Rot, Mr. Steckling attempts to squeeze the atmospheric blood out of the proverbial orbiting turnip - the moon, to prove that we should swallow the moon's habitability, at least along the termination line between the light and dark sides of the Moon. Why? He said, because *"the Moon's atmosphere, nevertheless, is dense enough to support clouds and vegetation"* (Steckling p.5).

Of course, anyone with any sense of investigation will find this book to be another example of outrageous synthetic selenography and lunatic geology. It supposedly proves alien settlement of the moon thousands of years ago. Steckling tries to follow up and prove the George Leonard theory, which he follows religiously, that there is plenty of evidence that the aliens never left, but are actually STILL digging for minerals, building space domes, and constructing underground cities! Fred

Steckling is another Moon Monument Mad Hatter mythomaniac!

According to Fred, we can no longer believe the text book propaganda that *"the Moon is incapable of supporting life, is airless and is, plainly speaking, a globe of dead rock."* For, he suggests further, that with a good magnifier glass and a thorough study of (blurry) NASA Lunar photographs, the diligent student (stooge, dunce, imbecile) will see the marvelous wonders, such as cities that grow, hundreds of artificial lunar domes, constructions and artifacts that cannot possibly be explained any other way than of alien origin.

Needless to say, grab your seats and prepare for an enlightening implosion into the imaginative cratered blunder-world of Fred Steckling's aliens on the Moon. We are going to see that the Moon is as alive as the earth – alive and full of pipe-dreams, phantasms, hobgoblins and witches on brooms! There is no doubt about it, if we take his word for it, that it is crawling and bustling with as much activity as Manhattan, New York, looking not too close with unbiased eyes, we will see that "It's Alive," and screaming of liveliness.

Fred would have us to believe, after pointing out references to centuries of lunar observations, that the Moon is filled with anomalous and mysterious "crisscross lines" (highways?) within the Crater Schroeter, "tube system" type constructions in Crater Cassendi, and twenty mile long "glass-lined" tunnels running between the Craters Gassendi and those of Messier.

The evidence does not stop here, but continues indefinitely with such examples of space alien manufactures as a black wall in the Crater Aristarchus (supposedly made of glass) that was not there before, and is not there now, and we may as well say was never there to begin with.

Without slamming on the breaks of our un-reasoning, we may Moon-cruise ahead and easily find "artificial mounds," "alien domes" about the Crater Archimedes, and other junk surrounding the local areas, along with various other mysterious geometrical shapes (Steckling p.12).

Furthermore, with no delay, we must orbit forward and keep our hats on tight as Fred blows more hot air artifacts up our exhaust pipes. For, besides these highways and byways, tunnels and funnels, there are more evidences to be obtained. Look closely and you will see UFO shaped objects, elongated cigar shaped crafts, digging machines, mining equipment, pleasure vehicles and lodge dwellings, weird carved symbols, signals, markings and best of all, old rivers and some few lakes, clouds, rivers and even vegetation!

In the Craters Plato, Aristarchus, Messier, Eudoxus and Mare

Crisium (not to mention, a million other craters), it can be seen (with not only a powerful telescope, but an strong imagination as well) mysterious moving lights, blazing triangles, flashing signal lights, long cigar-shaped alien craft, industrial platforms and other utility buildings (p.15-16).

Since the moon is alive and well (p.13), let us grab a shovel and straddle our brooms and take off to the moon to see for ourselves. But mind you, it will all be in old black & white, whereas, we shall enjoy the Technicolor of the imaginative interpretations.

Some of the first, and I mean the first of materials to ponder over in lunar selenography, are the locations about the area Oceanus Procellarum, an area between the Craters Seleucus and Hevelius. The Lunar Orbiter photographic equipment captured this area filled with alien debris and what looks like "ponds" as can be seen in (Steckling's choice of) NASA image IV-157-H3. The photo spans 30 degrees, from 30 degrees N X 55 degrees West, to O degrees X 55 degrees West, and exposes what Fred claims are UFO hangers and swimming pools, if not agricultural ponds.

Of course, these ponds could also be hydrogen fuel supply tanks, since UFO's run off of hydrogen power! And, because that equals "water," they could be alternately used to swim in on hot days, after heavy work schedules. They were obviously piping the water underground from such places as these lakes. Anyway, let us see what Mr Fred has to say, and then do some studies of our own for verification to determine if there are such structures on the moon.

1
ALIEN UFO HANGERS PONDS, CLOUDS AND OTHER MYSTERIOUS MOON BLOBS

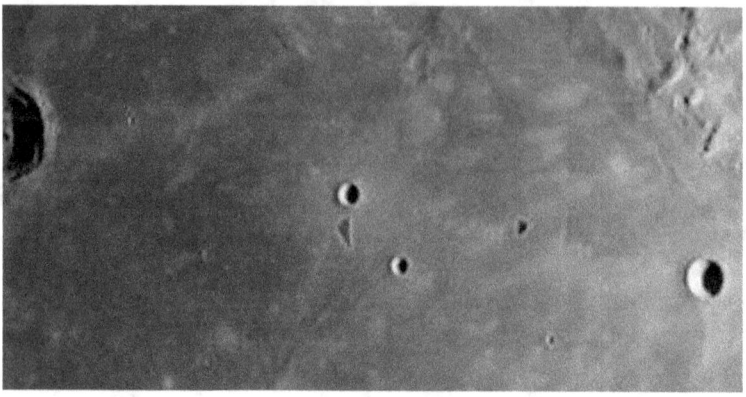

Lunar Orbiter photo IV-157-H3

Fred imagines he sees an artificial crater and a twenty five mile long artificial "UFO Hanger," a large structure supposedly used for hiding alien space craft. Fred said that "without a doubt, nature does not produce objects positioned like the well-formed crater and this perfectly straight mound or hanger, which is some 40km (25m) long." (p.85)

Fred's photo reference (157-H-1) on page 88 is NASA's Lunar Orbiter photo L.O. IV.157-H3 and 157M (Medium shot). The above photo shows Fred's unnamed crater and an alien "UFO hanger" nearby. Below are my additions of a "pond" and a crater. My selected pond is taken from one of those emulsion splatters, which we shall talk about soon.

Fred imagines he sees an artificial Crater and a twenty five mile long artificial "UFO Hanger", a large structure supposedly used for hiding Alien Space craft. Fred said that, *"without a doubt, nature does not produce objects positioned like the well-formed crater and this perfectly straight mound or hanger, which is some 40km (25m) long."* (p.85)

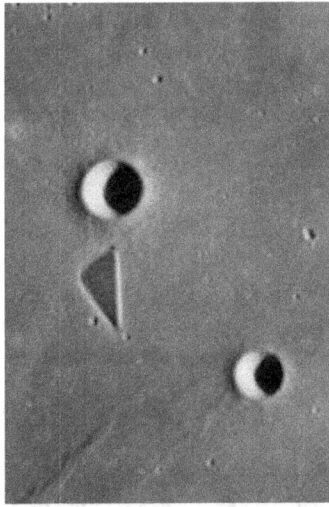

Steckling's enlarged and cropped image

Composite Photo

It is a weird photo with an odd shaped structure and is very impressive to the untrained eye. For those who trust Fred's reproductions in the book, the evidence is undeniable! With a little effort and some money to purchase NASA Lunar Orbiter Photo negatives, the student of alien artifacts can easily see that it is a farce!

The book uses horribly blurred and under developed photos. The reproductions are as bad as bad can get. With such photo copying anyone can take a NASA image and demonstrate anything he desires. And this is exactly what Fred Steckling did. For instance, here are some other examples of "blurry" clouds, giant finger-print like space bacteria, alien craft and other such nonsense.

Mr. Steckling knew exactly what he was doing and also knew exactly of his lying and forging "evidences" when it comes to supporting alien activity on the Moon. He was writing a book to sell and he had access to the best negatives NASA had, just as anyone of us today can have. All a person has to do is order the same materials from the NSSDC. (The National Space Science Data Center), which is the repository of most everything NASA has available on the subject of planetary photography and see for themselves that Fred is full of Lunar hot air! Never mind the men behind the curtains of NASA, who will argue the real truth, Fred has it all figured out for us! NASA is the one lying and covering up secret data and knowledge from the public.

In reading Fred's research on page 85, we are to believe that what is depicted here are alien objects, located at 55 degrees West by 25 degrees North. This is an area that covers some 319 by 81 miles of lunar surface! This location lies west of craters Aristarchus and Herodotus. This coverage is located in Oceanus Procellarum, east of the craters Seleucus, south to crater Hevelius.

On the NASA print one can easily see emulsion 'splatter' marks and development "blobs," and running completely off the image area into the non-image edges between the frames! Surly, without doubt, Fred was not blind nor was he ignorant of this, as to completely miss such a stark raving white emulsion splatter flying across his negative! It is unheard of short of forgery to misrepresent an obvious photo development flaw. This is perpetrated hoax.

2
MOON MICROBES

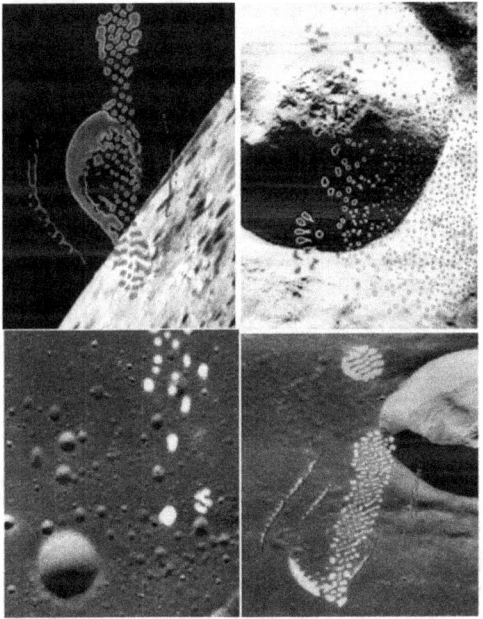

Alien UFO's, Lunar Clouds, Moon Microbes?

Mr Steckling's photo on page 88 and plate 61 is presented to prove without doubt that there is a 25-mile long "Hanger" structure adjacent to the Crater. Actually it shows no such artificial structure at all, but only a development flaw! So what is it? Is this alien evidence in the negatives or evidence of alien emulsion splatters on the negatives? Even with a two digit IQ it is not hard to see that these are development imperfections.

If we compare NASA's IV-157-H3 with other Atlases of the Moon photographically covering the same area, we will notice the missing emulsion splatters in the Crater Plato. Comparing the atlases with the NASA negative reveals just what we suspected and were looking for: The truth is the alien construction is not there, was never there and is not there now. The only place it can be found is literally "on" the NASA negative!

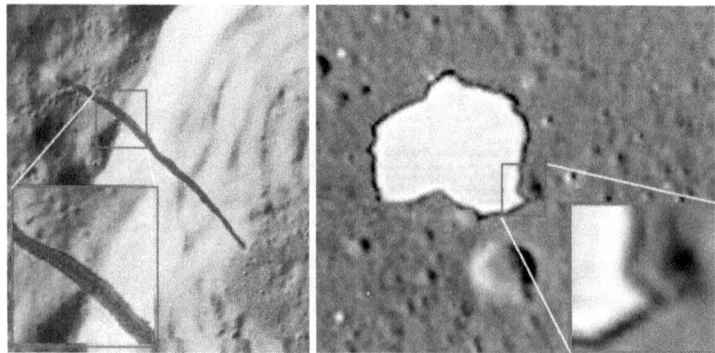

IV-128-H3 Water slide   III-194-m Pond or blemish?

Researching the atlases produced no substantial results in support of Steckling findings. For example, the so-called "alien hanger" or "pond" that appears to be alien in structure, after careful analysis was not alien at all. It turned out to be an emulsion artifact and is only alien to the NASA image.

Steckling presents one of the worse examples of the area. It is so blurred and printed with the dot matrix pattern in such a spoor fashion that anyone ignorantly trusting Steckling will obviously believe he sees some alien activity site. Even I was somewhat surprised of this before I saw the same image in full resolution clarity directly from the NASA print!

Further studies proved that this alien structure was no alien artifact after all. When the 157-H-3 Neg. from NASA was scanned and enlarged, the so-called "hanger" was revealed as an emulsion Blob! Just another water or moisture emulsion 'blotch'! Here are some more processing artifacts:

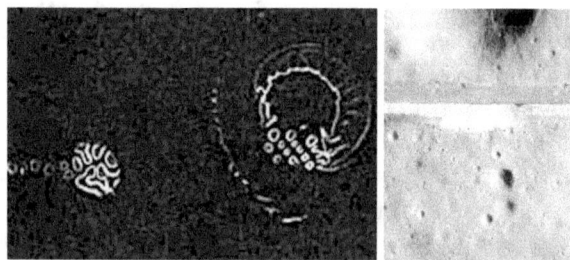

Space Bacteria?   LO-II-108-H1 Hair?

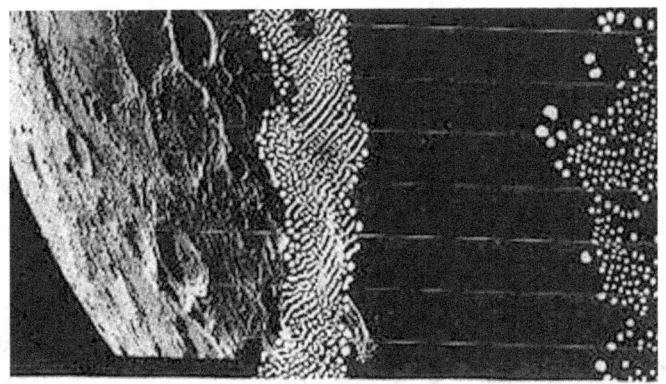

Colonies of space bacteria?

II-108-H1 Moon Worm? II-037-H1 Highlight?

3
ALIEN IRRIGATION PONDS

IV-128-H1

Throughout the Lunar Orbiter Photographic Atlas pictures one can find other similar alien structures. In the L.O. IV-128 there is a crater next to a larger crater or another "alien hanger" - depending on how one looks at it. This is one Mr Steckling seems to have over looked.

The warehouse hanger appears to be another one of those "ponds" according to Steckling, one that he likes to flaunt as alien in origin. Comparing the above frame to the larger Medium framed 157M the artifact hunter can see a larger section of the same area, but on a larger scale. Close inspection reveals that the alien hanger 'blob' is missing! The blob hanger disappeared! It appears in one frame and NOT the other! Do we have movable hangers? Hangers ¼ mile in size that are portable?

Where did it go? Well, it just moved somewhere else and we do not know how. Closer inspection reveals that there are hundreds of these 'hanger' blobs in many of the NASA negatives. The aliens must have airports, hangers and adjacent swimming pools everywhere on the Moon. 157-H-3 appears to have a few splatter blobs, and 157M also has TWO separate horizontal emulsion patterns. Comparing these emulsion blobs with the hanger reveals that it is a recurring stray emulsion splatter. It has something to do with the development system on the orbiter.

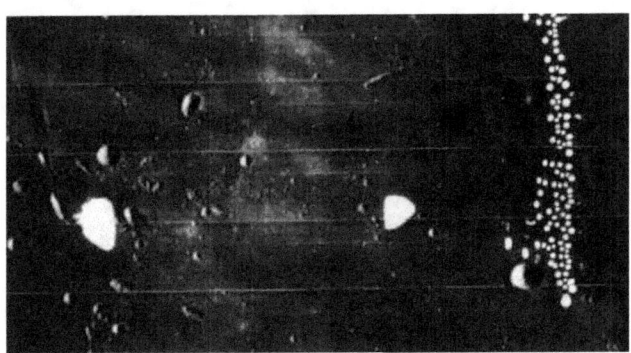

IV-114-H1
Steckling's "White Ponds"
Notice the supposed "clouds" to the right.

Another interesting feature about these blobs or alien hangers is that under magnification, enhancement and advanced focusing, the blob-hangers are more focused than the local lunar terrain surrounding them! This therefore shows us, by association, that the hanger or pond is just a BLOB!

A photograph is a photograph and each has a uniform focus. Nothing can be more focused than any other thing in the picture. Notice how the hanger sits fine focused upon the surface of the photo as an independent additional "thing" contrary to the less focused lunar surface features. Everything is slightly blurred except the blobs.

The same type of splatter blobs can also be found in Steckling's other "pond" photos in LO IV-128-H-3, which should actually be frame number "1." Fred said the structures are irrigation ponds, canals or dams. When enlarged the "ponds" display the same effect of being superimposed over one half of the raised lunar surface, which is also much less focused than the ponds! He said they are irrigation ponds situated among water bearing clefts and the rough terrain about the Crater Plato.

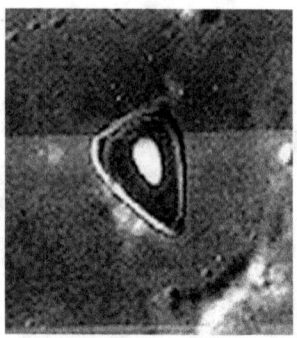

**LO-IV-128-H1 Close up.
Compare photos 128-1, 126-1, and 114-1**

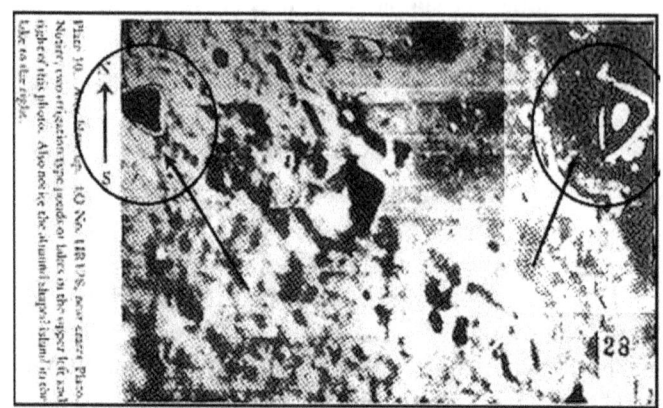

IRRIGATION PONDS

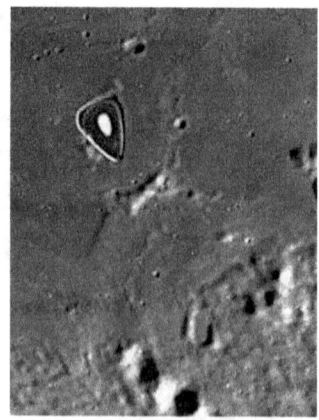

L.O. IV-128-H-3

In all cases the "ponds" and "hangers" display the same smooth rounded and triangular pattern. Compared closely with the true splatter patter blobs of 157M one can see that Steckling FABRICATED his "pond" and "hanger" theory! Such a wild imagination! What a best seller for UFO buffs to read! From such poor photo reproductions, they DO look like California irrigation ponds! (See Plates 55, 56, 60. Revised Ed., Plates 51, 52, 56)

It appears without contest, that in ALL cases the ponds and hanger images display the same finer focus with microscopic jittery edges, just as we see in the true emulsion splatter blobs in 157M photo of Crater Pluto.

So, now we know "WHAT" these blob hangers and ponds are. But, can we find more of these 'ponds' that Fred missed? Are there more to be found in the frame "1" series of photos? Yes there are plenty of them. Here is a list of the NASA Lunar Orbiter Photographic frames (L.O. IV "H-1" series) showing this "same" exact set of Ponds [1].

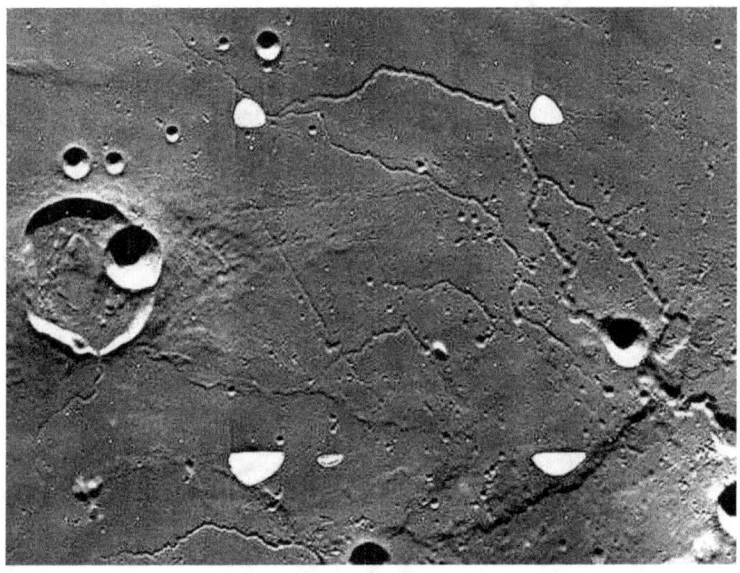

IV-151-H1
Steckling Plates 57, 58, 59, 57A

What is also interesting is the symmetry involved. It is as if all the High resolution Frame Number 1's seems to always have this splattered pond effect. The emulsion appears to re-appear over and over again. Apparently, in developing the H-1 frames they must have experienced the same "consecutive" development problem. Many even run off the edges of the photos! The consecutive examples in the Lunar Atlas show that they are a series of splatters unique to the #1 series of the frames.

In checking the Lunar Orbiter Atlas page by page for other "ponds" reveals that there are over 57+ plates out of 404 photos that have these anomalies or MULTIPLE POND emulsion blobs. Many have splatters and dotted 'cloud' looking patterns, and fingerprint patterns next to them, as the photos above show (See 114h1 photo).

Out of 404 photos every frame number "1" has the same TWO ponds. There are white ones over black areas and white ones over visible terrain (transparent ponds). Also clouds, which are also emulsion splatters with the fingerprint looking clouds sometimes accompanying them. After careful reviewing of all 57 plus plates and their supposed multiple "ponds", I found a single common denominator. They ALL show exactly the same number, distance and symmetry.

Compare the following set of "ponds." Look at the geometric layout of the 'White' Ponds! All have that slight triangle look with the same axis line. See my drawing about this phenomenon. What is interesting is all the axis lines have the same perspective points. The duplicated symmetry is staggering and apparently ALIEN according to Fred! Alien, because he didn't choose to reveal this repeated pattern in all the Lunar Atlas photos to his readers. I presume that he choose NOT to because it would have exposed him as a fraud!

Comparisons made of all the listed examples indicate that the ponds (the TWO ponds that always show up in every photos listed), are the SAME TWO emulsion blobs, which show up exactly the same way in ALL the photos listed! It is the same with the FOUR pond sets. This obviously presents us with one conclusion. That these sets of "ponds" are emulsion splatters from photo development processes.

MOON MARS MONUMENTS MADNESS

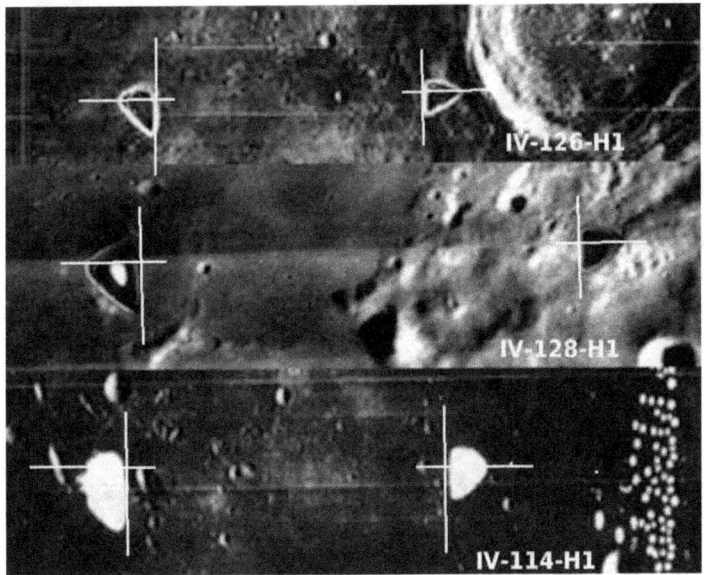

Emulsion Blob "pond" symmetry

When NASA photo 128-H1, is enlarged, what do we see? We see the same 'pond' anomalies as found in the previous 157-H-1 photo. Fred argues the same asinine rhetoric as before to prove his case: *"The most revealing things on this picture are the two irrigation type ponds or lakes with reinforced walls. They are triangular shaped on the left side of the photo, the other on the right side, with an almond shaped island."* (p.82)

Fred asked a question here, *"Is it perhaps a pump station?"* Well, NO, not really. Fred could have given us the excuse that these ponds must have put on some alien theatrical and moved as quickly as the Lunar Orbiter Camera moved to shoot them. Of course, taking a fixed position in front of the camera as the camera scanned the regions.

There are simple explanations for these technical problems:
1.) The "ponds" are emulsion blobs in the development of the negatives. 2.) They are specks of dust on the camera lens. 3.) They are some form of light leakage into the camera that appears on all the frames numbered "H-1" They do not appear on frames H2 or H3 or any other ones.

In one photo the pond is miraculously transparent showing the fault line under the pond! Why would aliens dig a pond right through a hill?

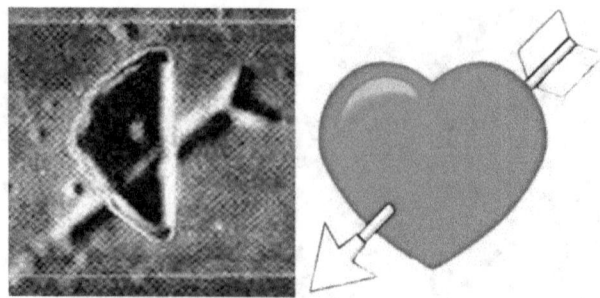

IV-139-H1

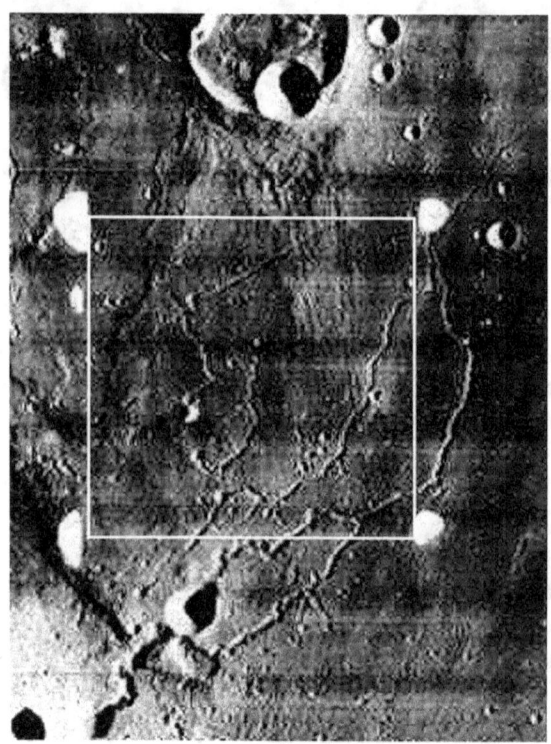

151-H-1
Duplication of the emulsion blobs show the ponds to all be the SAME set. It is the same with the set of four ponds.

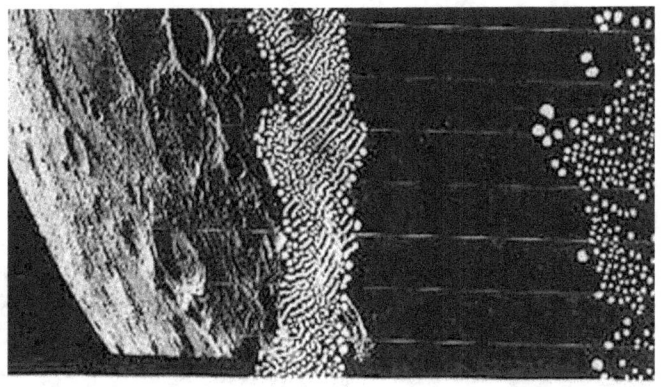

Check out these bacterial 'emulsion' splatters.

Fred uses the photo 151-H3 of Crater Krieger and the surrounding areas to prove they are advanced alien constructions. (P.101 Plate 53) There are four ponds as above shown. He said, *"To the right of the crater Krieger, notice five triangular shaped irrigation ponds, or reservoirs just like the ones on the previous pictures. Except here, the sun shines from the north just at the proper angle and altitude to reflect its light off the water, which shows up white in a black and white photo."* (p.114)

Computer enhancement reveals these are photo development errors and are not found on the lunar surface. The blob is sharper than the blurred Lunar surface, which contradicts LO photographic focus consistency. The ponds have more clarity than the lunar surface, which means to me, that the camera did not shot them from off the lunar surface at all, but must have picked up this added phenomenon from inside the camera or during the development of the negatives. Again, they only appear on the high resolution frames "H1" and not the Medium resolution shots.

4
CLOUDS OVER CRATER VITELLO
L0-V-168 M

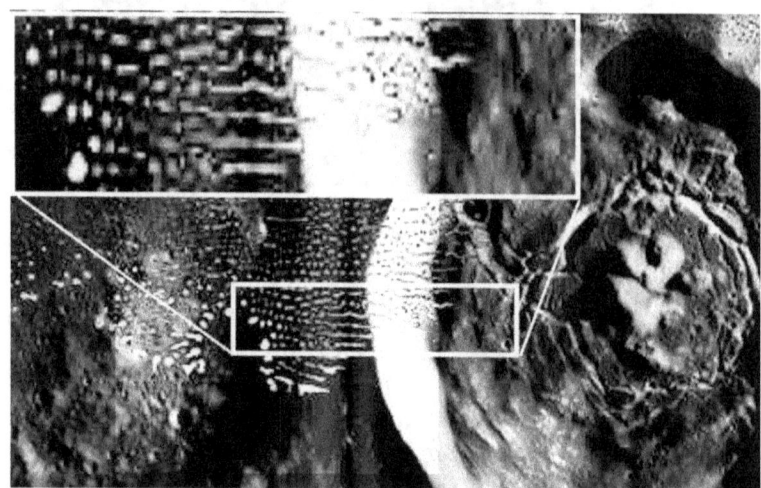

Notice that the "clouds" run toward the right side off the lunar surface!

Fred's next bold attempt to establish an alternative Lunar Atmospheric Science is with LOV-168-M), his photo 45, page 77. Crater Vitello is shown here with "clouds" floating over the sky above and across the Crater Rim. Fred said, *"Notice cirrus type clouds left of crater rim. Also notice platform in the center of the crater."* (p.77)

Cirrus type or Macheral type clouds? What a fantastic scientific observation to throw off the ignorant lunar meteorologist. HOLY MACHERAL! This photo is fishy to me! A real fisherman's tale, too!

All Fred gives us here is a "chopped" photo. Now what difference does this make? Is this just a cropped version to fit the book for publishing? NOPE! He crops the picture to hide a very important exposure (pun intended) of his fakery.

A little investigation into the complete NASA negative reveals that these "clouds" are miraculous clouds and not Macheral clouds, and that they are not clouds at all, but is obvious emulsion blotches, blobs, splats and splatters. These so-called clouds "miraculously" float right off the edge of the negative and into the dark unexposed area of the film proving it is all a development flaw. There are hundreds of them on the Lunar Orbiter photos!

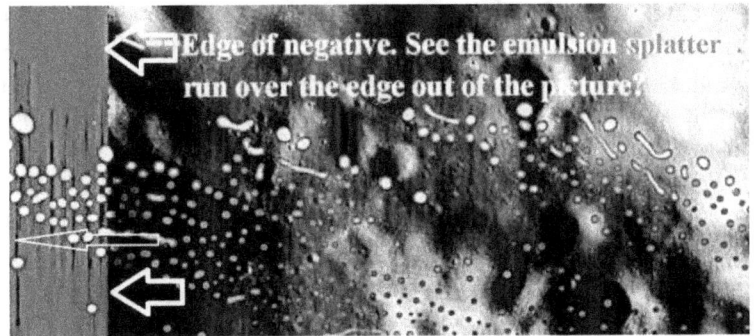

Emulsion splatter runs off edge of picture.

Now, why the lie? Why the big bold-faced misrepresentation of the facts and the hiding of what the negative really shows? Fred said he had access to them! Why did he assume they were clouds when they are blatant emulsion problems? A BEST SELLER! Hell man, it was the early 1980's and everyone was into UFO's and Aliens, just as they still are today! And so, THE BOOK! "We Discovered Alien Bases on the Moon."

Well, we have discovered alien interpretations of the Moon alien to what the negatives really reveal. The Lunar Orbiter V photos show that there are at least 18 such "cloud" photos and probably more if one takes the time to search for them. [2]

Why are there so many selenography researchers hell-bent to lie to the public? Answer: To promote the 'Alien Agenda" and hide the boring truth, discredit religion and real science, propagate hyped up sensational misinformation and promote book publishing. Not everyone in the 1980's could afford to buy up all the NASA negatives to verify the truth. Maybe the public does not want the truth and they prefer to want aliens to exist. Maybe they WANT to believe nonsense and mythologies because it's fun, easy to believe and does not demand responsibility. Maybe they want fraudulent soothsaying evidence to pacify their fears of a silent universe. Maybe they are not happy with a simple God and simple religion of hope and prefer a more romantic story that aliens are coming to save them from atomic destruction. Maybe people are just stupid, uneducated and gullible! It is the opinion of many decent scientists that the alien presence myth offers an alternative hope evasive of human responsibility and void of future judgment, and that it is a replacement for traditional theology.

Human nature gets real bored with the mundane reality of life, a life void of mysteries, suspense and the magical. There is a part of human nature that loves the horrible, the outrageous, alien invasions, threats and doomsday scenarios. Myth brings relief to daily drudgery, work and utilitarianism. Science fiction seems to release stress and open doors to what is not natural to human existence. The fantastic is necessary if there is no true god to turn to. The extraterrestrial alien savior is the new cosmological answer to the human dilemma. Nevertheless, some of the public may actually be interested to know that they are being lied to, if they could see the truth about these photographs.

The truth can be stranger than fiction. Moreover, what is stranger than this moon monuments mad fiction is the fact that the majority has swallowed the E.T. hook line and sinker. It would appear that human nature prefers the lie over the truth, for the truth always brings human responsibility.

It is easy to see to see how people can be deceived. But to lie and deceive the public for self-gain is intolerable! The only theory is "THEY" can do it and get away with it with immunity. They, as well as the government can also make a lot of money at it too.

Nevertheless, there are those who disapprove of such deception, if for anything, for the sake of the truth. Hence, the purpose of this "Moon Monuments Madness" book is to expose the fake alien habitation of our solar system.

5.

CRATER DOMES AND ALIEN HOUSING PROJECTS

Now, what else have these myth-makers found? CRATER DOMES, little lunar houses for aliens to hide in, weird work worm-hole lava tubes and meaningless mining operations have been some of the many mythical inventions fabricated by the spin doctors.

LO-IV-187-H2
Take a look at this "dome" sitting on the edge of a crater.

Mr Steckling again has found some more interesting alien artifacts in the Orientale basin area. About the highly fractured hummocky terrain there are numerous odd shaped domed mounds, some of which have fractures on their crests. Fred thinks these are alien structures and finds what appears (and I mean "appears") to be a mound or blob sitting on half of a crater.

One of the most interesting "domes" found by Fred (and every other NASA geologist) is located at 85 degrees W X 15 N. Mr. Steckling points out that these pictures also show examples of atmospheric clouds and alien cigar shaped vehicles.

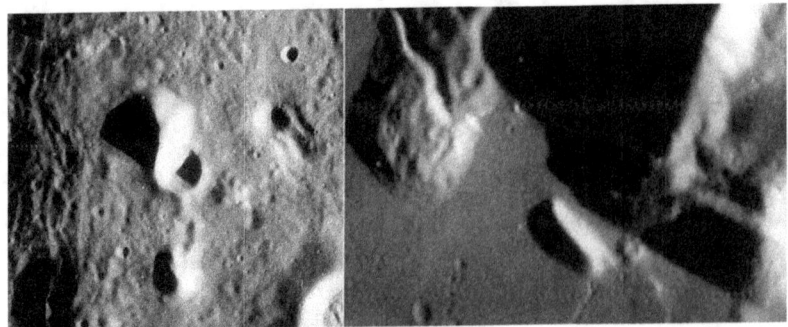

IV-187-H2

The crater mound "dome" and another cylinder shaped dome.

In Plate 66, he said that, in *"The Alpine Valley area of the Moon... [We may] Notice (a) large cloud-like object covering half of the central crater."* In Plate 67, Fred insists that in an area blow-up of LO IV-187-H2, we will undoubtedly notice an *"oval shaped object, UFO or cloud, one mile east of the large crater.* And that we will also see *"More clouds are in the upper right plus three domes (two northwest of the crater)."*

These so called lunar domes are only examples of natural lunar morphological processes. Geologically speaking, domes can be naturally made. These natural mounds show up in other locations in the Lunar Orbiter photos. They are common in this area and are part of an inner blocky ring encircling the basin at the same radius. This particular IV-187-H2 *"crater mound"* (above) indicates an isolated peak surrounded by or protruding through this unit. (Shultz. p. 478). The crater rim "dome" has at least one other possible explanation besides the crater ejecta theory. It is the "fumarole" or gas vent theory.

In Plate 67a, which is LO IV Photo 187-H2, he continues to suggest that *"we can see a large cigar shaped object above the crater floor."* (p. 120-121).

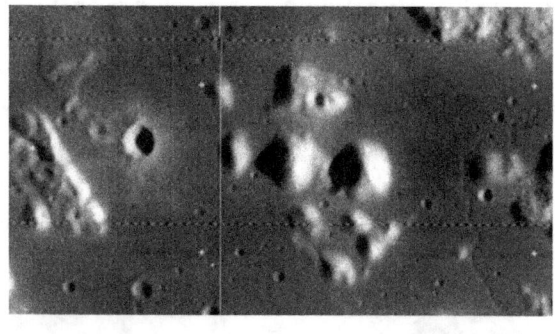

Other "clouds" or "domes" found in this area.

A close-up reveals no alien activity around the site to suggest that the mound is 'artificial'. Actually, the area is less chaotic than other locations and should not be a surprise to a lunar geologist.

The following illustrations demonstrate how an impacting meteorite can strike the lunar surface causing volcanic disturbances, release gas "fumaroles" and spew out lunar ejecta regolith materials, thus creating a lunar regolith mound to pile up on one side. Photo IV-187-H2 admittedly presents an interesting geological event, yet nevertheless, a naturally created one.

A fumarole (Latin *fumus*, smoke) is an opening in a planet's crust, often in the neighborhood of volcanoes, which emits steam and gases such as carbon dioxide, sulfur dioxide, hydrogen chloride, and hydrogen sulfide. Fumaroles may occur along tiny cracks or long fissures, in chaotic clusters or fields, and on the surfaces of lava flows and thick deposits of pyroclastic flows. A fumarole field is an area of thermal springs and gas vents where magma or hot igneous rocks at shallow depth are releasing gases or interacting with groundwater. The formation called Home Plate at Gusev Crater on Mars which was examined by the Mars Exploration Rover (MER) Spirit is suspected to be the eroded remains of an ancient and extinct fumarole. [R.V.Morris, S.W. Squyres, et al. "The Hydrothermal System at Home Plate in Gusev Crater, Mars". *Lunar & Planetary Science* XXXIX(2008)].

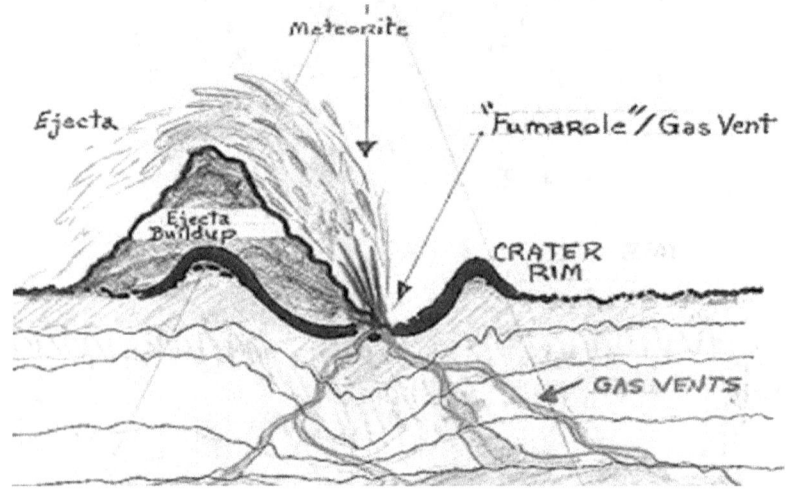

Fumarole Gas Venting

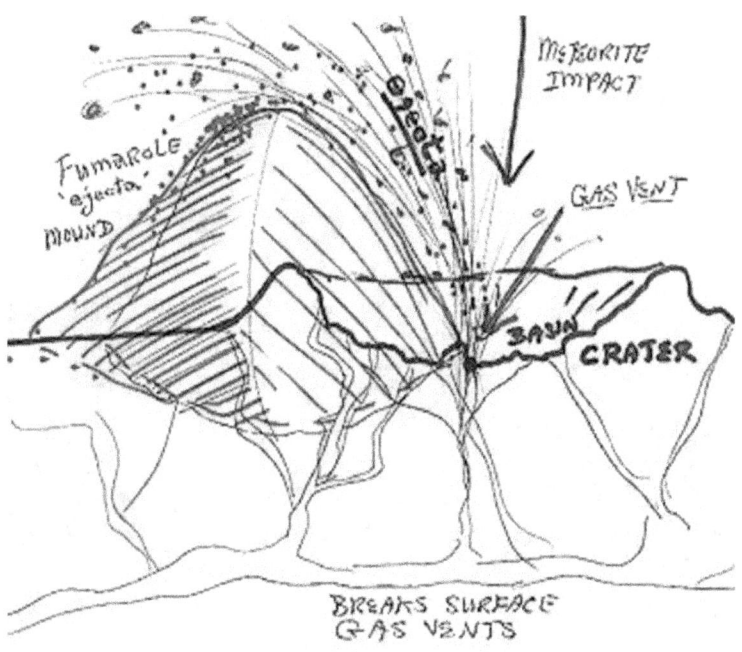

6
FRED'S DAM'D CRATER WALLS AND RILLE RIVERS

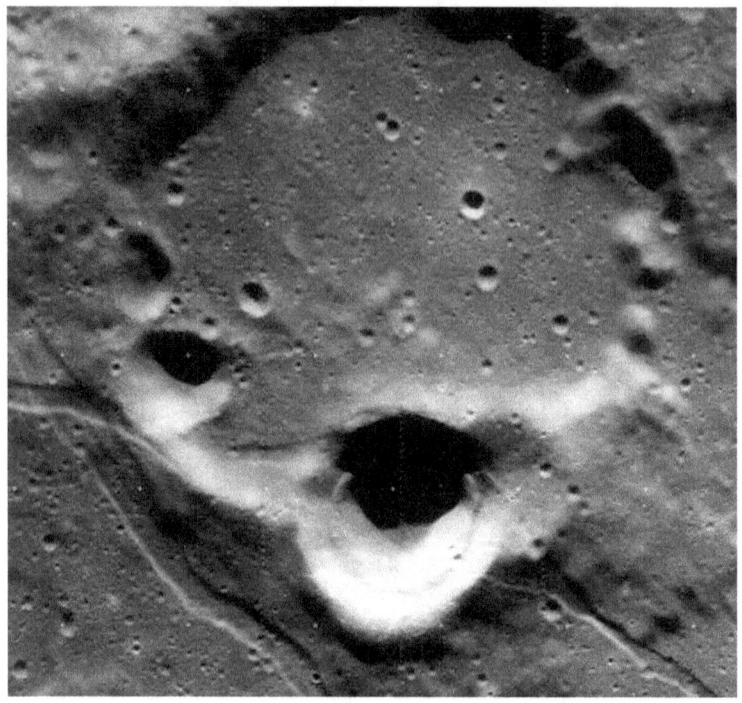

LO-IV-161-H3
Close-up of the crater with dam walls.
(Plates 71 and 72, p. 126-127)

Steckling's problem of wrongly numbering his NASA negatives seems to be an on-going problem. This misled me for a while, for about six months, looking aimlessly for this location on the wrong frame. The Lunar Orbiter IV photos gave me a much clearer picture of the surrounding areas. I found the correct frame and also found no need to waste any more time or money on NASA Negatives. I found the problem!

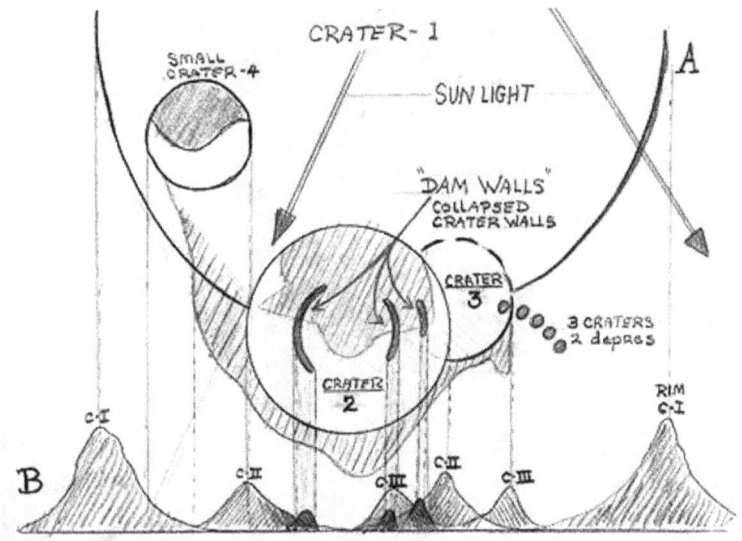

CRATER DEVELOPMENT

Fred said, that when the Moon had water, the aliens converted Damoiseau-D Crater (161-H-3) into a gigantic reservoir by building what 'appears' to be dam-like walls inside the crater. Again, Steckling misses the frame number calling it IV-161-H1, while it is actually the high resolution frame H3. This is the crater Damoiseau-D.

Fred imagines he sees a "smoke pillar" drifting from west to east, clouds floating over the crater's edge, a long platform entering a hanger, various assorted constructions, platforms and a larger platform extending over a riverbed (p.102). Besides all this selenology silliness, he sees two more double walls, more riverbeds, what looks like dam walls and five evenly spaced constructions, *"much like pump stations"* (p.103).

Well, I admit there are many signs of erosion in the area, but does erosion necessitate alien activity, artificial dams, drainage canals and pumping stations? Of course not. All of these things can be explained by natural lunar geological and morphological processes.

All of Fred's photos, including this one, are blurry and undecipherable. No one can really tell much about the details in his book and I am sure he knew this! One's imagination can run wild with blurry photos. The same goes with cloud formations. People see animals in cloud formations all the time. But, in the latter case, sane people know that there are no real animals in the clouds except for air plane pilots, birds and high flying insects and maybe a hail-storm of frogs.

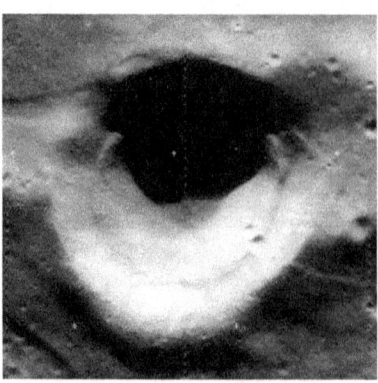

Enlargement of crater Damoiseau D

Once released, the untrained imagination, prompted by the whorish ramblings of Mr Steckling will humbly follow and believe anything. Imagination derives from "image" or 'icon', which is the basis of 'idolatry', a false representation. With this in mind, anyone can see how a person reading this cheap balderdash can contract all kinds of adulterated ideas and absurdities flying through their mind.

So, Fred has found, in (Plate 70), many proofs to alien activity. He has smoke pillars, clouds, platforms, constructions, rivers, pump stations and dams. After reviewing frame IV-161-H3, it will be found to be true: The "smoke Pillar" is an old double crater rim highlight from intense sun light. The clouds over the crater rim is sun light at same angle as inside the opposite wall of the crater.

The platforms and hangers are more intense sun highlights burning their way around and over the lunar hills or from the photographic glare when the picture was taken. The constructions are crater rim ejecta, highlighted and shadowed with light playing on the surface. The riverbeds may very well be some sort of ancient flow either by water, lava or volcanic erosion and surface fracturing. The pump Stations or five objects described by Fred are three craters and two mounds or depressions casting shadows. The dam walls inside the crater are two opposing crater wall collapses. Other craters show this kind of double crater rims. One is usually larger with the smaller one are inside the larger.

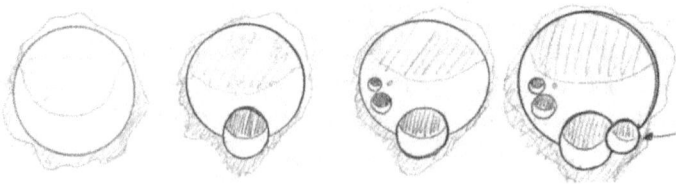

C.　　　D.　　　E.　　　F.

The crater's formation from multiple impacts

C. Older original crater

D. Smaller secondary impact crater

E. Other minor impact craters

F. More impacts

COMPLETE CRATER LAYOUT DRAWING

Notice the progression or history of the development of the crater through crater impacts. Crater C is the original crater. Crater D adds a secondary crater that sits on the previous crater rim. E and F add other minor craters overlapping the original crater. These multiple crater impacts cause crater rim collapses, thus creating what Fred calls dam walls inside the crater.

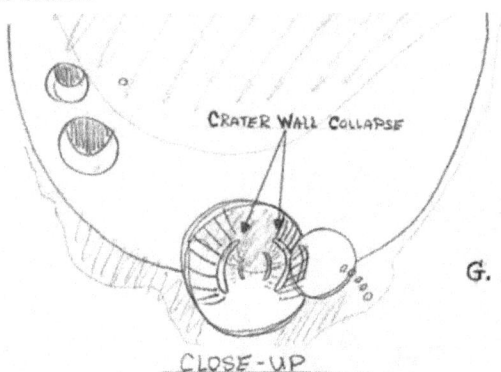

Crater wall collapse

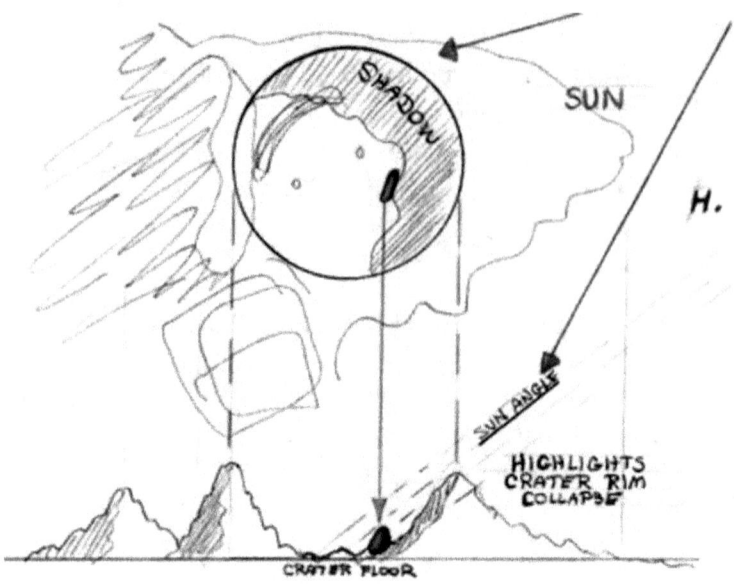

The bright "cigar" shape is now apparently just another chunk of collapsed crater Rim. Here is what the crater geology is showing.

7
CIGARS ON THE MOON

Craters with hidden UFO's? Cigar shaped alien space craft? This is what Fred sees in this particular NASA frame. In the center of Crater Romer there hides an alien UFO. As Mr Steckling said, *"Notice the large platform and long object on the center mountain."* (p.161).

AIRFORCE BOMBER FOUND in a Moon crater?

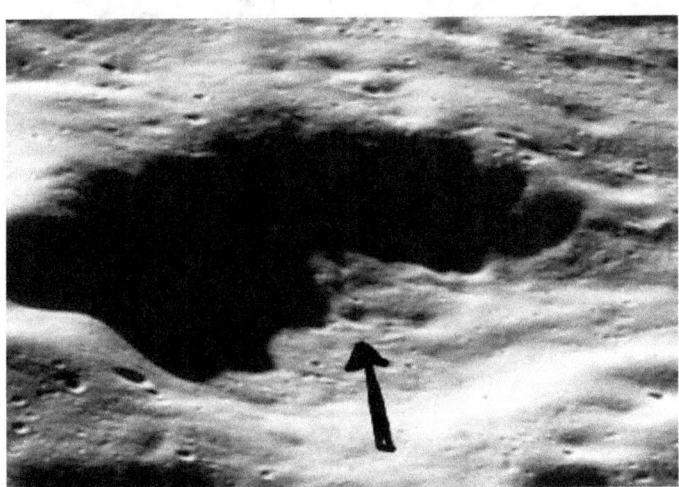

AS16-118-18923   A Crater Cigar!

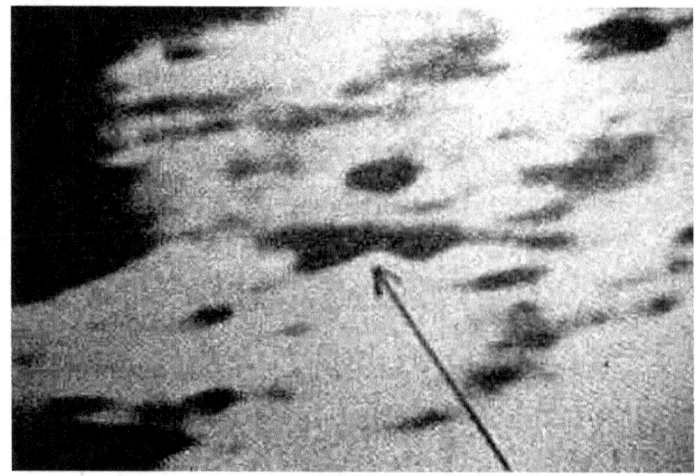

AS16-118-18923
Fred's Close-up of "Cigar UFO"

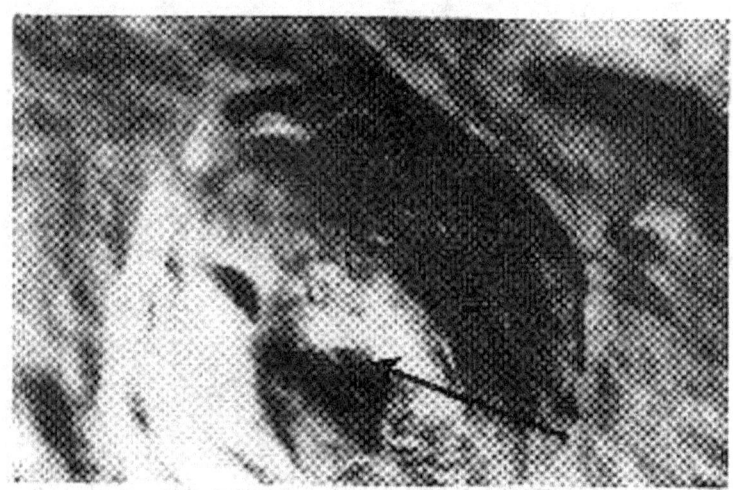

Plate 110. LO IV Photo No. LO IV 89H3. The Crater Romer. Notice large platform and long object on center mountain.

IV-73-H3
Plate 110. The Crater Romer: Notice the large platform and long object on the center mountain.

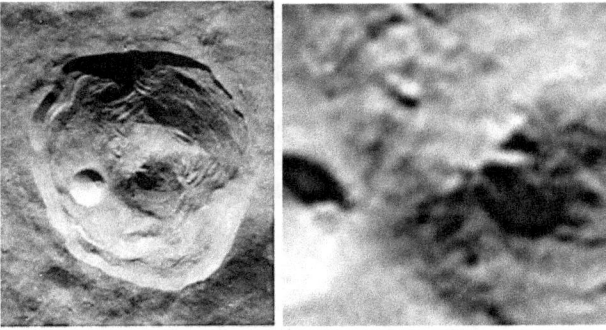

Fred lists Crater Romer as on Lunar Photo IV-089-H3 and locates the alien space craft on the central peak (arrow Y). But Crater Romer is located in the Lunar Orbiter photo IV-073-H3. Photo 89-H3 lists three craters: Crater Alfraganus, Delambre and Zoller. Crater Romer is NOT located on this frame. These craters too have lots of central debris in them. Almost every crater has some of kind of odd boulder arrangements and central pile of dirt.

Crater Romer surely has debris in it as the picture shows. But alien artifacts? If left up to this kind of scholarship, every crater has UFO's in them.

Returning to the original location, the UFO surly looks convincing! Yet, when on closer examination, rather than a little air plane or UFO it is actually a pile of crater debris. In the center of the crater is a pile of boulders and rocks forming a single cone-shaped central peak that actually may have a summit pit?

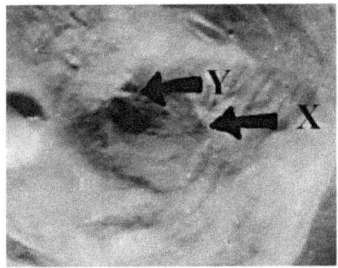

A closer cleaner picture of Romer. Fred's Ref:
L.O. IV 089-H3 Should be frame IV-073-H3

**Another view of IV-089-H3
and other crater examples of interior debris.**

Shultz said (p. 100), *"The peal comprises the entire crater floor and crests 1.8 km above its base to the east* [i.e. between X and Z along the crest] *but is about 2 km below the eastern crater rim. Thus the three-dimensional view has been greatly enhanced. Adjacent to its west slope is a ringed depression (arrow X) that might represent a collapsed form."* This collapsed area can be seen within the whitened area where the two arrows are pointing. At "Z" is where Fred sees a UFO.

On closer inspection, the little UFO is actually some over-hanging rocks that receive a highlight from the sun while below is the shadow area.

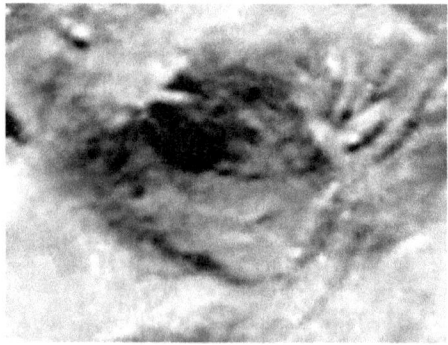

Overhanging highlight area seen in these Apollo frames.

AS17-P-2293

Other crater "cigars" and alien oddities are found in almost all of Fred's photos, as if alien spaceships are just lying around all over the lunar surface. Maybe they are just rocks, boulders and crater ejecta?

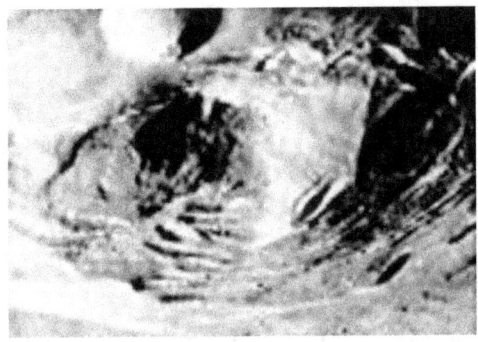

AS17-P-2295

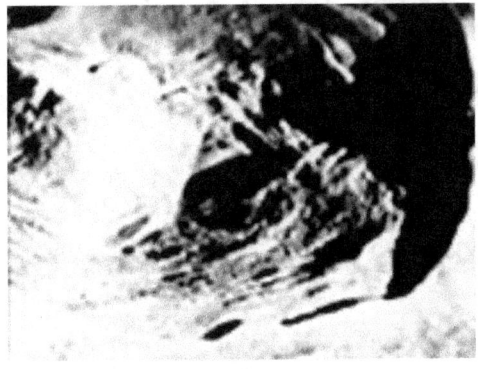

AS17-P-2298

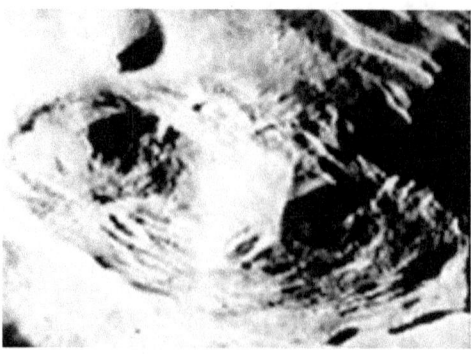

AS17-P-2300

Try to find other cigar shaped items if you can. Fred found them! We have shown a few (below) that are very obvious and one is used for the front cover of Mr Steckling's book.

AS16-19238
Looks like part of the Orbiter?
Background to Fred's front Cover

8
ALIEN LISTENING DEVICES

IV-169-H1

Bull's Eye! Fresh new crater inside older crater!

Besides crater walls, collapsed rims, and flying cigars notice Mr. Steckling's alien made crater device. He said, "*See the double crater in the larger Humboldt Crater. This one looks artificial, Mining or*? Well what is it? He never says what it is actually. Maybe it is just a double crater!

Of course, Fred would have us consider that they are giant alien speakers or listening devices similar to the Arecibo Observatory. Here are some other craters that depict speaker-like shapes. Unlike craters formed by endogenetic (volcanic) processes, double craters may be formed by rare coincidences of a meteorite striking in the middle of an older crater. This is implied by the subdued nature of the outer wall compared with the fresh "crisp" sharp edge rim of the central feature. See IV-169-H1 photo frame.

Geologists tell us these weird looking double craters are examples of natural lunar phenomenon, where crater rims have collapsed down inside. Another possibility is that two meteorites struck the same spot one after the other in the same way that multiple meteorites strike successively in a row forming crater chains. Still another is that a new meteorite struck an older one as implied by the newness of the inner crater rim compared to the older outer crater rim. Yes, they look weird, alien and artificially created, yet they are only the product of natural crater processes.

Fresh crater impacts in area of older surface in area of Fred's double crater

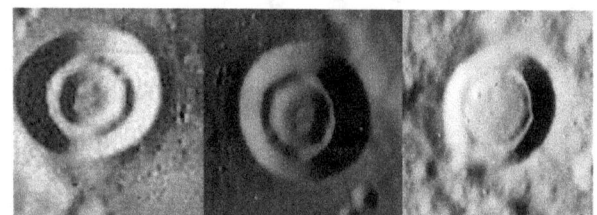

IV-125-H1   IV-119-H3   IV-156-H3

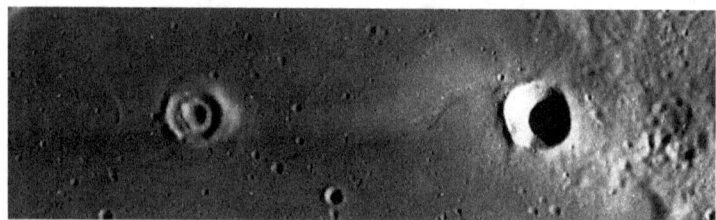

IV-131-H3   Other double craters

9
PLANETARY PIES
THE "PIE-CUT" MOUND

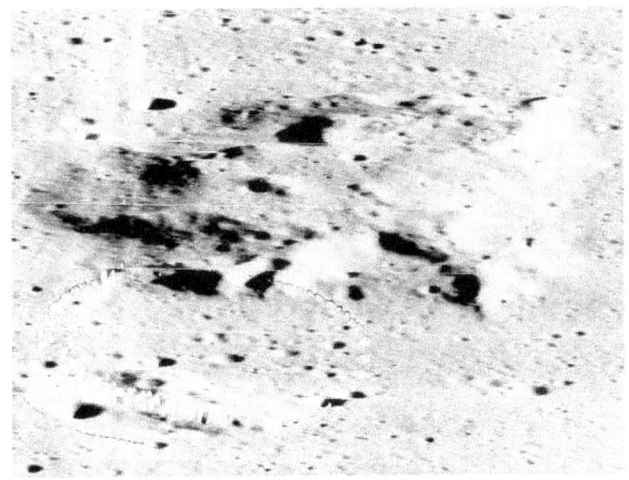

LO-II-213-M and H2

Lunar Orbiter II photo 213-M is a very beautiful and romantic lunarscape shot. This is a great site that excites the wondering imagination. This is Steckling's "Pie-Cut" mound site. His most accurate pie-hole said, that the *"hill, left of center, has been cut out like a piece of cake. Above the hill appears large construction casting a shadow to the left. To the right notice several perfectly cut holes or craters and two oval objects on the rim."*

After careful study of the original frames, there will be found no constructions casting shadows but for some shadowed craters, no perfectly cut drilled holes accept fresh crater impacts and no oval objects on any crater rims, though one can see a play of highlights against shadows all over the frame.

Also, one can see little white equidistant dots dividing the scanned area. Some of these white dots hit exactly on some crater rims causing what looks like UFO's and "alien objects." If one takes the complete frame and follows the white dots, they proceed all the way up the frame and into dark space, perfectly divided according to the distance between the scans. Wow, what a "piece of cake!"

Pie shaped cuts or landslides play a common role in lunar surface features. So do cracks, gas vents, double and triple craters, crater chains, multi-rim craters and extremely odd shaped boulders. This pie-cut is nothing other than a collapsed mound. At closer range in a good zoom in on the mound, one can see that the pie-cut thief has left some residue at the inner edge.

The Lunar Orbiter Atlas "The Moon as Viewed by Lunar Orbiter," has a large reproduction of this pie-cut on page 122. A drawing will show this land slide effect we are talking about.

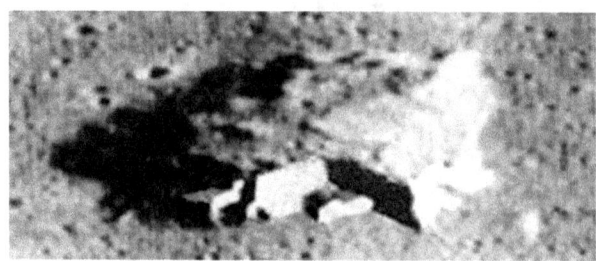

The left side appears collapsed and debris has slid down, somewhat obscured by the heavy sunlight hitting it. The right side is in shadow. Collapsing mounds do not always create rounded openings, edges and cliffs, etc. This is what makes this formation unique.

The composition of the mound determines the differences. In this case, the mound could have been struck by an early meteorite and then affected this way by lunar erosion. Microscopic meteorite bombardment over long periods of time could have caused secondary erosion patterns, such as the pie-cut effect. Even adjacent land flow could have pulled away and removed debris. The apparently visible erosion effects as seen in 213's medium shot and in 213-H2 evidence this.

Fred's perfectly cut pie hole and the two oval objects can be easily explained, without having to assume alien influences. Heavy shadow and light play a major role in such odd looking land masses as this pie cut. Shadow and light playing off round boulders! Yes, boulders can be round! Even square and rectangular! A good whack by a small meteorite can crack one even into a triangle shaped rock!

"PIE MOUND"

The reason the crater holes look like the ends of large tubes sticking out of the lunar surface is because they are in extreme light and shadow. The rim's details are obscured either by excessive shadow, which the Moon has plenty of or by highlights that tend to bleach out the details on the sunlit sides in the photos. All that would be left in this case would be the sharp rims that have no delineation as to the oblique crater walls and regolith soil.

Under more balanced lighting conditions and by a different camera and sun angle, the crater(s) would look the same as all the millions of other craters!

The funny white "finger print" image is a development process disfigurement. At about 10:00 O'clock, Northwest in the photo is actually another pie cut mound, yet not as visible as Fred's mound. There is land flow around the area with newer crater impacts.

There are at least five large craters and numerous smaller ones within the pie-cut area. There is much lunar regolith disturbance. Now that we have shut the pie-hole of Fred's alien pie-cut mound blab and have shown it to be nothing but a mound of dirt, we may proceed to Fred's bowl of alphabet pottage soup and other supposed alien objects.

10
CRATER CAMBELL ALPHABET SOUP!

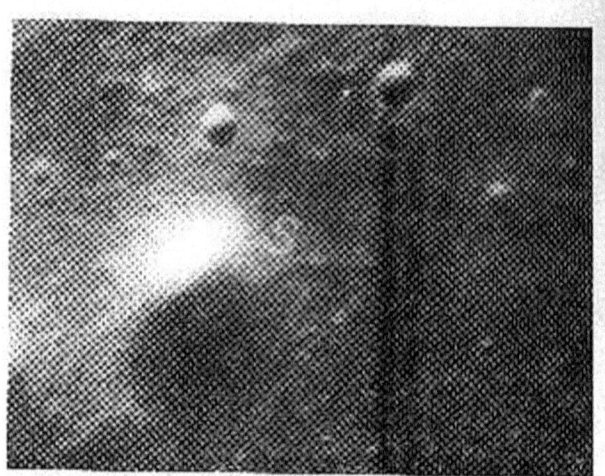

Plate 109. Apollo 14 photo No. 14-80-10439. Another letter "S". This time placed beside the crater.

AS14-80-10439

This is an example of one of Fred's "S" Letters, which he finds scattered all over the moon inside craters. This time, this "S" is not found in a crater. It is sitting next to a mound. This is a familiar "emulsion hair" caught in development of photo AS14-80-10439, which by the way cannot be located. It is not on the internet and is not in the NASA Apollo-14 photo archives. It is said to be classified? It is nowhere to be found. In fact, I cannot locate film Magazines 79 or 80. See footnotes of reference of image close to this one proving magazine 80 does exist [1]. Maybe someone has these sets of photos?

In L.O. III photo frame 194-3, Fred sees the alphabetical letter "S" artificially inscribed within a crater. It looks like some 'signature' or locator marking for bypassing UFO's! Admittedly, this really excites the human eye and throws a real S-curve ball when first viewed. This appears to be a real genuine anomaly. Needless to say, this is not some artificially created mark, but just another explainable lunar phenomenon.

A viewer will see the same effect when looking at a river from an airplane. Many rivers curve and meander like the coils of a snake in an S-shaped pattern. Well, so do crater rims when they collapse. Each end of the "S", the bottom and top being crater rims remain. The pushing in of the two collapsed rim creates the middle of the "S". They look connected and probably are from the erosion collapse of the rim. Notice that the entire rim does not have to collapse, just two opposite sides. Other crater rim materials can easily slide down creating the connecting parts of the so-called "S" shaped letter.

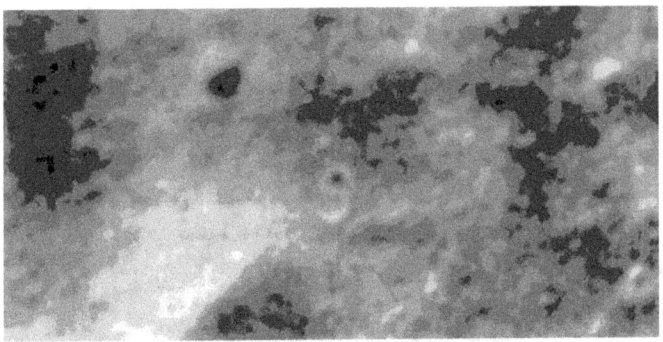

Another shot of AS14-80-10439
(http://keithlaney.net/ApolloOrbitalimages/AS14http://keithlaney.net/ApolloOrbitalimages/AS14)

III-194-H3
Fred's loony alphabetical letter "S"
Plate 120, p.141

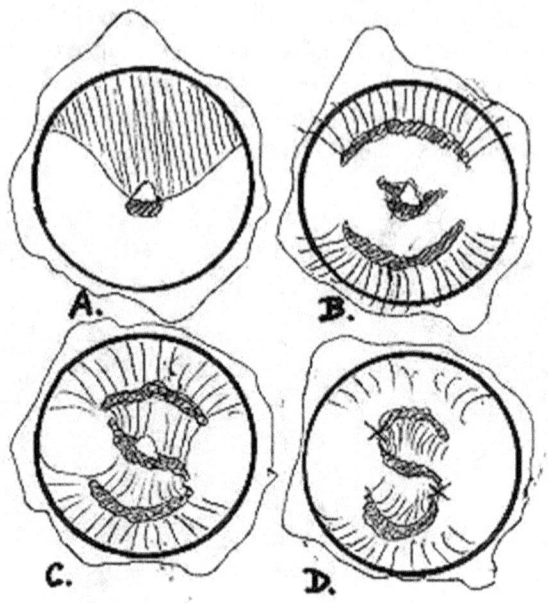

Explanation of "S" shape development inside craters.

If Fred would have studied other crater interiors, he would have to assume aliens were learning the English alphabet or something! Crater floor features can appear to look like many things to alien artifact hunters. When you add numerous geological events together, such as crater rim collapses, double impact craters, double impact crater rim collapses, boulders, cracks, fractures, shadows and highlights you can have aliens trying to rewrite Shakespeare!

Many craters have this "letter" effect phenomenon. After craters are formed, regolith sometimes, and in most cases, pours back into the crater from the rims. This creates mounds and interior rim ridges, scarps and other funny looking piles of "twists and turns" that would appear from a distance to be "letters." Some blurriness does not hurt in adding more grammar to the effect.

Fred found other consonants such as the letter "R" half-hidden in a crater shadow on the back side of the Moon. See p.140, plate 119 and plate 122.

The Letter "S" seems to pop up in many craters, usually smaller ones. Why so many small craters and not the bigger ones? Why not crater Plato? How about Crater Tycho? The drawings we have just seen are a series of examples of how an "S" can be formed inside a crater.

The impact leaves a small mound or ridge in the middle of the crater.

Crater rims symmetrically slide inwards as they collapse causing the top and bottom of the "S" to form. The collapsing debris forms uplift in soil forming the middle of the "S" and a ridge connecting the top and bottom of the "S."

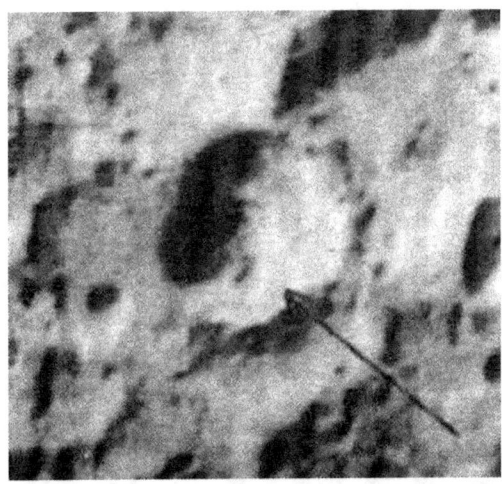

(Fred does not give us a photo number for this frame)

Using fuzzy versions of Lunar Orbiter photos one cannot help but see what one wants to see. Yet, there are details enough through crater comparison to illustrate how shapes like this can form. With slightly blurred images, shadows and highlights illusions emerge causing many factious features to appear. If the details are smoothed out around the regolith mounds the sliding rims may become anything the alien artifact hunter wants them to be. The three parts of the crater collapse, the top, bottom and middle would tend to blend together under such lighting. Exposure also helps to blend the disconnected material into a perfect looking "S" shape. Thus, according to these effects some look complete, while others appear fragmented.

According to Fred's photo references from Apollo-14-80-10439, which shows a white letter "S," many NASA photos do present white streaks, hairs and dust scratches that create what might be misinterpreted as alien artifacts. They are so numerous that one might suppose that E.T's once inhabited the whole surface of the Moon. Hotels, motels, houses and mining operations seem to be everywhere according to this man's theory.

For example, just enlarge any Lunar Orbiter photo and POOF! You will have thousands of alien structures, buildings and craters with alphabet soup letters! The "S" in plate-121 is obviously a curvilinear shaped dust "hair" caught in the development of the photo.

How about BLACK letters? Can anyone find a BLACK letter? These would be harder to find, but not impossible to explain though. They would be cracks, fault lines, gas vents or even small rilles.

11
TANKS AND TOWERS
Alien Oil Refineries!

AS16-118-18918

Another, strictly speaking, Steckling hideout for alien structures is NASA Apollo-16 photo AS16-118-18918. Fred said, fuel tanks, towers and other platforms are supposedly visible in this photo. Let us take a closer look at this.

The photo is very blurred and the highlights are too intense, while the shadows are very dark and imposing upon the terrain. With a super imagination, a Stecklingite can easily discern artificial structures with very little effort. What one stuffs into the pie is what one gets out of it after it is bake in the oven of warped imagination. Fred apparently sees geometric and other artificial shapes created by some super technical mind, as the following fantasy drawing shows. (See drawings 1 and 2). Here is his oil refinery tanks.

Another concept drawing of Towers

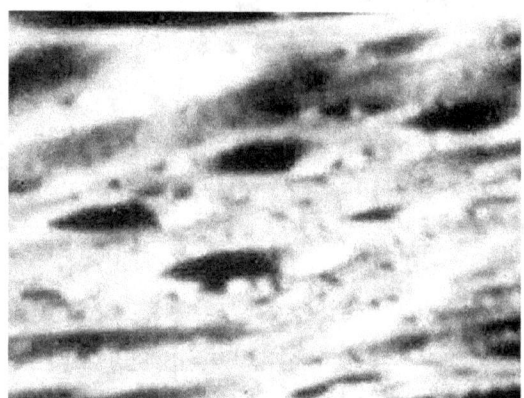
1.) Enlarged and enhanced

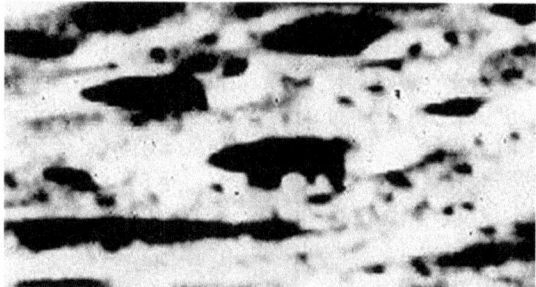
2.) Contrast added

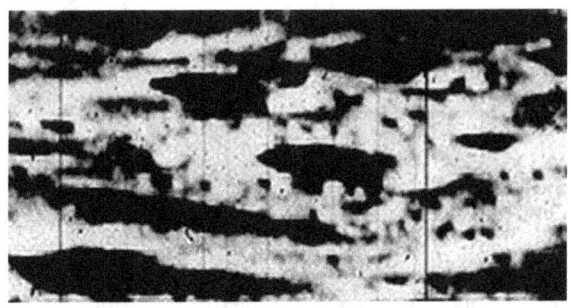

**3.) Sharpened
Fred's Plates 148 and 149 (Plates 133 and 134)**

Drawing-1 See the oil tanks?

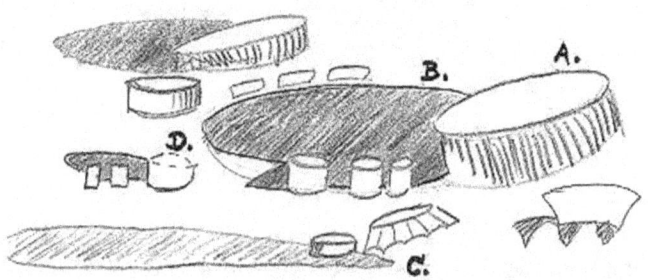

Drawing-2

    Contrary to such fancy interpretation one can easily extrapolate from the photo a purely 'natural' set of lunar surface features (See drawing -1) and twist them into "tanks," "towers" and "platforms." Whereas, after looking at the photo closely, with some enhancements, drawing – 2 exposes the "A" object as not a platform at all, but the Sun baked highlight of the right side of Crater "B."

    Object "B" is the termination line between the crater shadow cast by the sun upon the crater rim. Object "A" is the highlighted right side of the rim. "B" is the shadowed area. The crater to the upper left shows this light-shadow phenomenon much better, as do all the craters in this enlargement.

The little white 'geometric' obstructions around the crater rim are nothing more than what was left of the rim after it collapsed into the inside and of whatever crater ejecta was thrown out from the impact. They appear to be crater ejecta boulders. In some cases, the crater rims have collapsed, leaving "toothed" openings along the edges, which appear in blurred imaging as supposed artificial "digs."

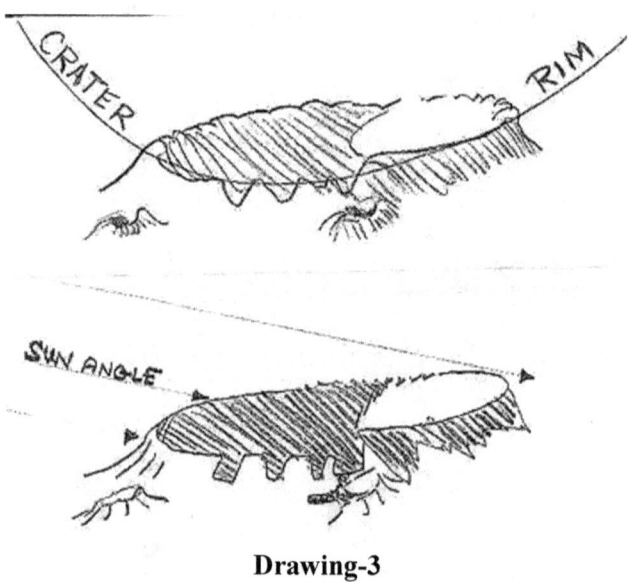

Drawing-3

Large chunks (not regolith soil) of rocks, boulders and larger pieces of the lunar surface (that was not broken up in the impact) might cause the toothed appearance. These are some that never loosened in the collapsing of the rim.

The openings are the ones that slide down into the crater leaving 'holes' or toothed ridge areas along the rim. Crater shadows would show through, while the sun would continue to highlight the remaining crater rim producing what appears to be 'teeth.'

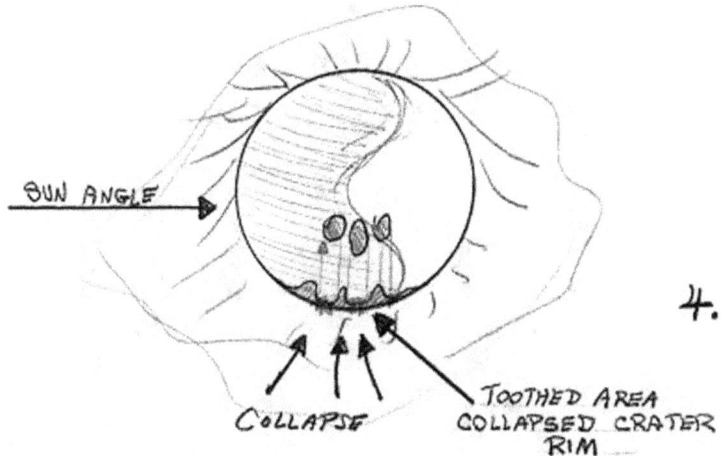

Drawing-4
Tanks, Platforms and Artificial Structures?
No! Simple crater anomalies.

12
ROLLING BOULDERS ALIEN MINING MACHINES

AS16-119-19067

Using alien mining operations as a premise, Fred argues for an artificial origin of these 'rolling stones'. He said the paths left behind the boulders are "tracks." These tread marks are the evidence for alien mining machines left behind after a great alien exodus from the Moon! Why they left no one knows.

Fred shows us some lunar photographs of (boulder) tracks produced by rolling rocks in the Crater Vitello as well as from other areas, insisting that *"Tracks of vehicles rolling over the ground, up and down the hills, have been photographed."* And as far as he can tell, the *"large vehicles, some seventy-five feet across, seem to probe the soil for future mining possibilities. The tracks left behind by these vehicles show definite 'stitch-marks' by some form of belted vehicle."*

The Apollo-16 photo 119-19067 shows a 30 mile stretch of stitched path or track left by some URO (unidentified rolling object) that tumbles along up and down the hills, and up and over the crater rim continuing down into the other side. Fred believes it must be alien because the tracks are *"identical to Plate No. 23 tracks"* (in V-168-H2). He forgets to explain how they appear out of nowhere.

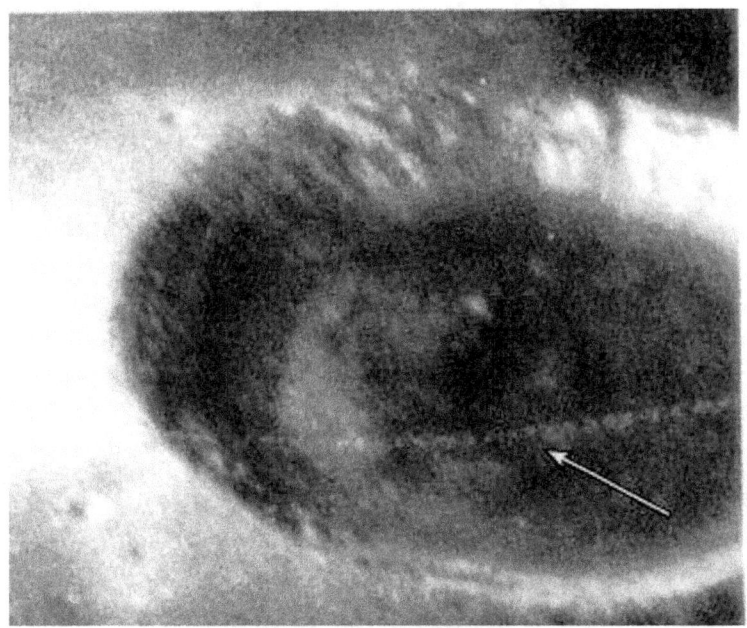

AS16-119-19067
Close-up Lunar back-side, Crater Vitello
(Fred's photo, Plate 24, p.39)

He continues to tread on our reasoning that one rolling object (boulder) measures 11 by 21m and judging from its shadow, is 11m tall. A second mechanical monster shows as a bright spot above the first one. (LO V-168-H2 "30.4 S, 38.5 W.) The large object leaves a definite 'stitch-mark' path caused by some form of belted vehicle (Plate 23, p. 39). The following some different shots of 168-H2. After close inspection the tractor looks nothing like a wobbling moon machine. What is seen are just boulders!

The second set of photos are of a famous boulder from the old Lunar Orbiter V shown in the 1972 research article "Boulder Tracks on the Moon and Earth" by H. J. Moore.

The track is produced by a skidding boulder in the crater Petavius. The boulder is about 68 by 81m; the track is 50 by 137m. (Frame H-36, framelet 144, 24.8 S, 59.8 E). The frame to the right is from the LRO Camera, frame No. **m183131379rc_pyrd1**.

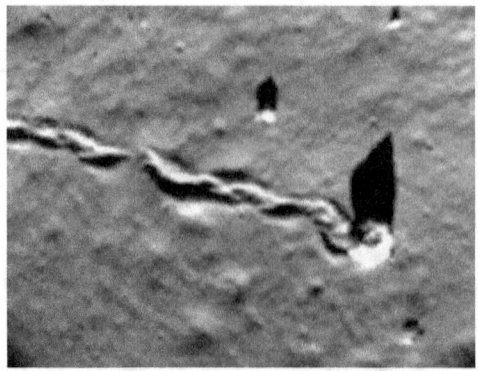
Cute little "rolling stone"

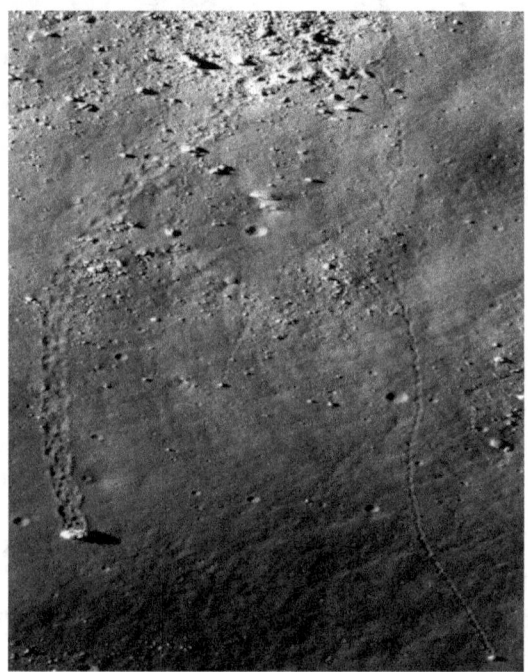
Photo: m135433752lc_pyrred
Lunar Reconnaissance Orbiter Camera photo

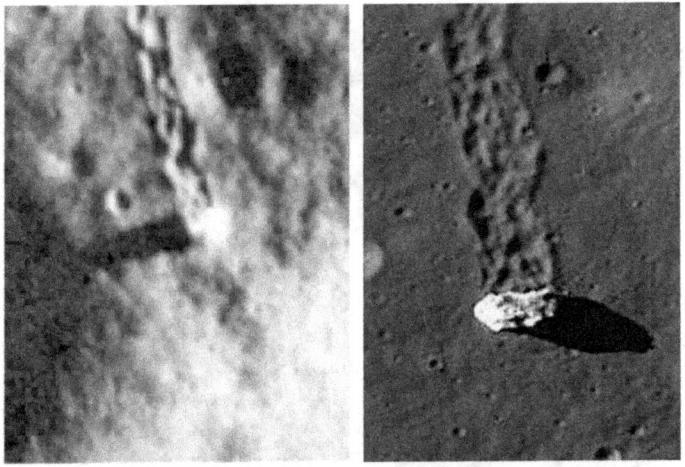

Older Lunar Orbiter V-168-H2 (Left) compared to the new Lunar Reconnaissance Orbiter photograph above (Right)

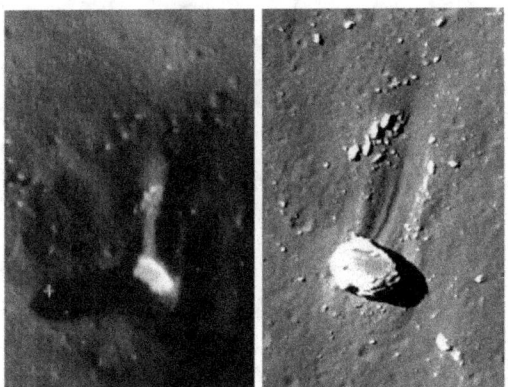

LO-V-036-H2 compared with LROC

This next track example was produced by a "walking" boulder in Scheoter's Valley. The boulder is 19 by 35m. The length of the track is almost 540m. Notice the zigzag pattern of the track and angularity of the boulder. It is located at 49.8 degrees West and 25.6 degrees North (Moore p. B171).

Fred said these tracks are "stitches" left behind by some alien "stitching" or "digging" machine. In his book, he sketches what he believes it looks like. Another Japanese web site extrapolates another design of what is causing the tracks.

LO-V-204-H3-[2]

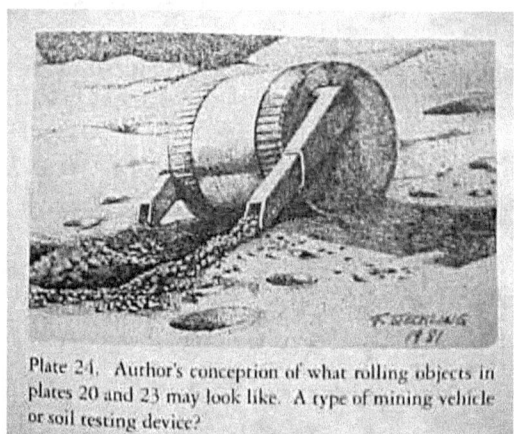

Plate 24. Author's conception of what rolling objects in plates 20 and 23 may look like. A type of mining vehicle or soil testing device?

Fred's drawing of the stitching machine

Fred must "dig" inventing alternative theories, purposely defrauding his innocent readers and defying all his so-called government training by believed this 'bull' dozer theory himself. Knowing Fred and knowing the NASA materials as I do, he looked at the same photos and negatives of the traveling clouds of the Vitello area that I have and probably disregarded any further study into the anomalies.

With a bit of effort one can read all they want about the truth of traveling boulders in Moore's monograph, "BOULDER TRACKS ON THE MOON AND EARTH", in USGS Research, 1972; Survey Professional Paper 800-B, pages B165-B174.

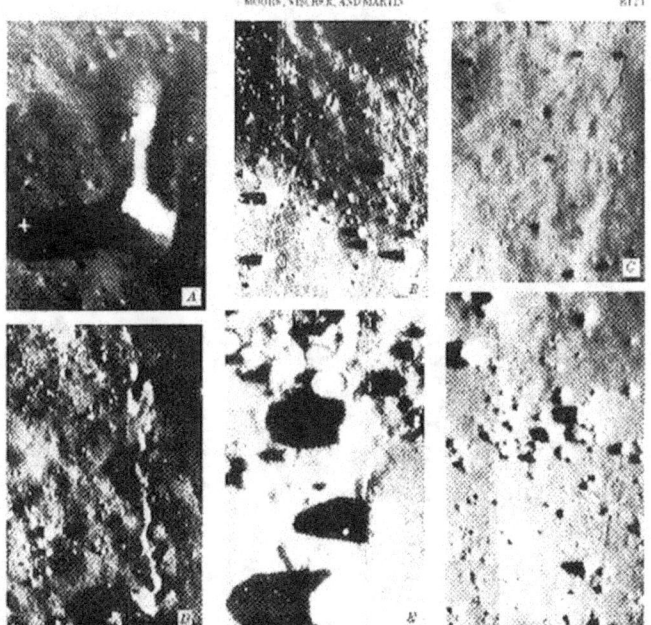

Geological Survey Research Professional Paper 800-B, p.B171.
Fig. 4. – L. O. V photos of boulder tracks showing types of tracks

In this study are examples of lunar and earth boulders rolling up and down hills. Other examples that I found are as follows.

LO-V-063-H2

Another taken from L.O.V-036-H3-[2]

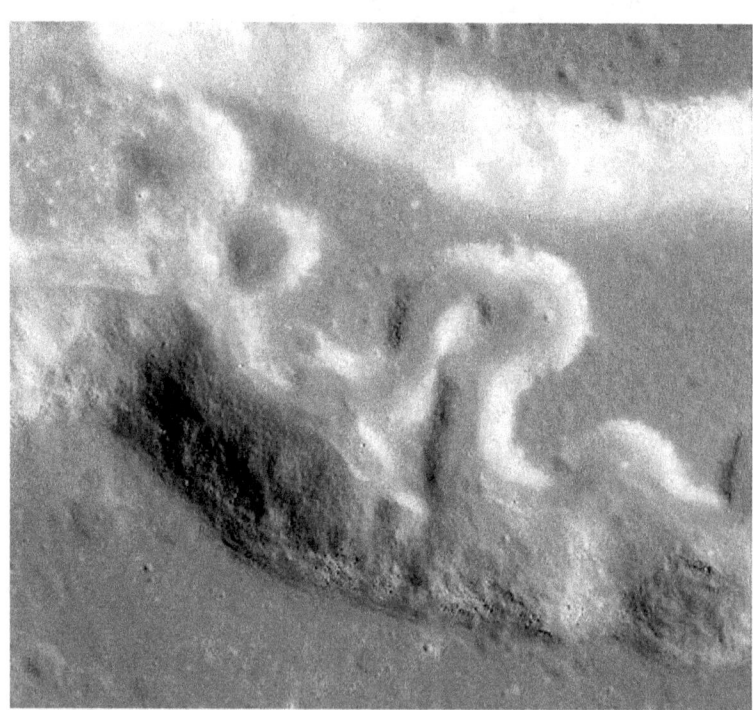

This image was captured by Lunar Reconnaissance Orbiter on August 14 at 00:18 UTC

The above picture is from Vallis Schröteri, the largest rille on the

Moon, and is on the Aristarchus Plateau. The channel actually contains two rilles, a larger one 4.3 kilometers wide and 155 kilometers long, and a narrower, inner one 600 meters wide and 204 kilometers long. The main feature is called "The Cobra's head".

Notice the long track going down the sloping wall of the outer rille. It is continuous at the top of the slope, dis-continuous toward the base. Is this the track left by a boulder that first rolled, then, as it picked up speed, bounced down the slope?

If Fred knew about Moore's article, he would have also have known they were ejecta rocks thrown out by crater impacts or some other lunar geological event. The above three writers in this monograph tell us exactly what they are and how they have done what we see in the above "traveling boulder" shots.

Here is a good shot of two boulders chugging along on the Lunar surface taken by the Apollo-17 astronauts using a telephoto lens to photograph "boulders and boulder tracks" on the North Massif at the their landing site. The largest object, making a weaving, tread-like path is about 5 meters across.

There are other lunar rolling rocks that Mr Steckling seems to have missed. Over 300 "boulder tracks" have been identified in photography taken from the lunar orbit. (See, Grolier, Moore, Martin in "Lunar Rock Tracks," 1968; and "The Moon - Boulder Tracks and Nature of Lunar Soil," 1973 by D. Reidal Pub.).

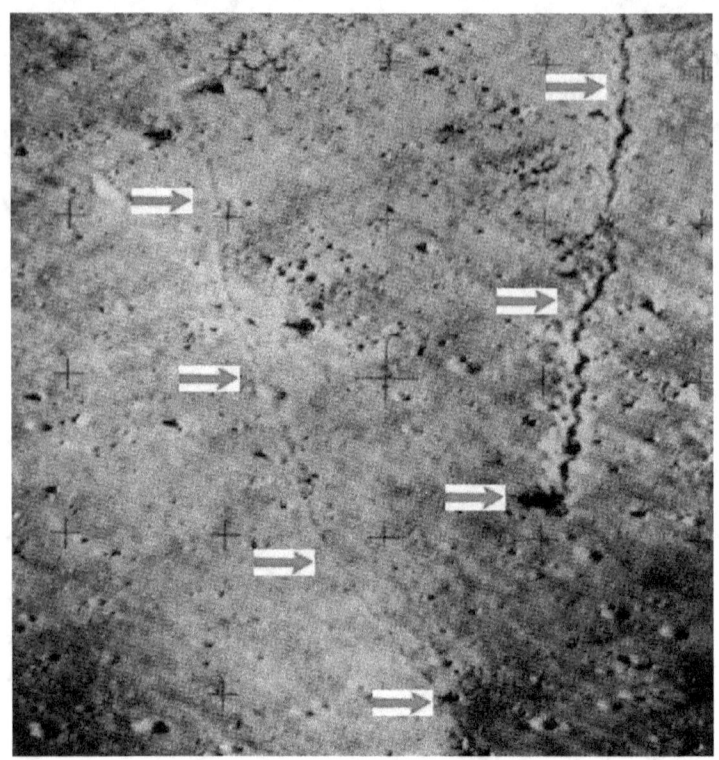

Apollo-17 photo AS17-144-22129 (H)

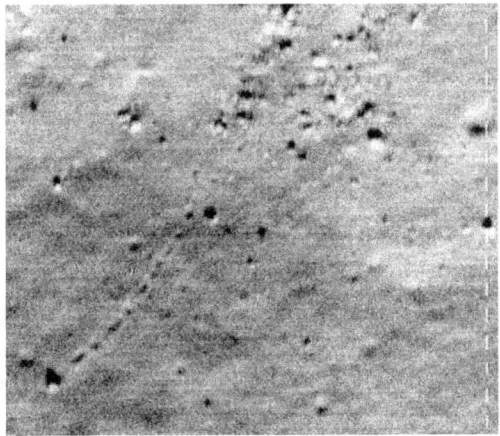

LO II-092-H1

The above frame is from Lunar Orbiter II at an altitude of 44 miles. The image is taken from frame 92, frame let 445, and has resolution is 0.98 meters/pixel. As such the large boulder that has left a trail is around 6-7 meters in diameter.

Right: LO V-204-H3 Photo of same rolling boulder as captured by LROC m188572067rc_pyrred (left). For exploratory fun in locating other Lunar "rolling stones" and "boulder tracks" see this web site:

http://the-moon.wikispaces.com/Boulder+tracks]

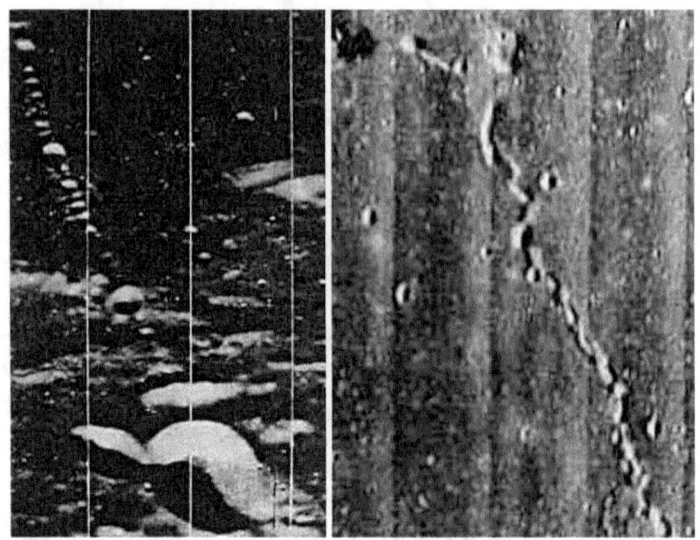
Moon as Viewed by Lunar Orbiter p.97, No. 119

13
CRATER LAKES

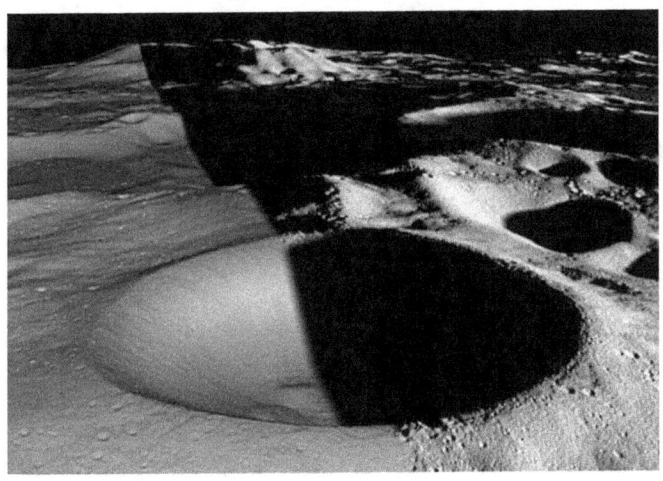

Is there water on the Moon? Fred Steckling said there was and he was right, but not to the extent he dreamed of. In a recent finding NASA said ice may reside in cold dark crater pockets, especially at the North and South poles of the Moon.

Actually, a team of NASA and university scientists using laser light from LROC laser altimeter examined the floor of Shackleton crater. They found the crater's floor is brighter than those of other nearby craters, which is consistent with the presence of small amounts of ice.

The crater is named after the Antarctic explorer Ernest Shackleton, and is two miles deep and more than 12 miles wide. Like several craters at the moon's south pole, the small tilt of the lunar spin axis means Shackleton crater's interior is permanently dark and therefore extremely cold. (Eddie Wrenn, "More Water on the Moon." [Ref. www.dailymail.co.uk
/sciencetech/article-2162505]

The South Pole is not the only place where water might be located, but the north Polar Regions offer equal hope.

Shackleton, (12.5 mile-diameter) permanently shadowed crater near lunar South Pole,. Possible an ice coating at its base

The North Polar Region is said to contain scattered pockets of ice and water, and a lot more than previously calculated. One estimate by NASA is that there is some 600 million metric tons concealed away in about 40 craters. This is not much, compared to Fred's estimate of "oceans" of water, but enough to supply a large city the size of Seattle for about three years. Water appears to be turning up everywhere on the moon. Not bad considering it was once thought to be bone dry. The latest discovery from the Indian Chandrayaan-1 lunar orbiter found 40 craters each containing water ice at least 2 meters deep.

Unfortunately, these small traces of water on the moon do not mean we are to start buying up water-front crater properties. Nevertheless, Mr. Steckling believes there is huge amounts of water, and not just water, but plants, trees, vegetables, lichens and mosses, and with all the dark spots at the poles, let's not forget the possibility of mushrooms!

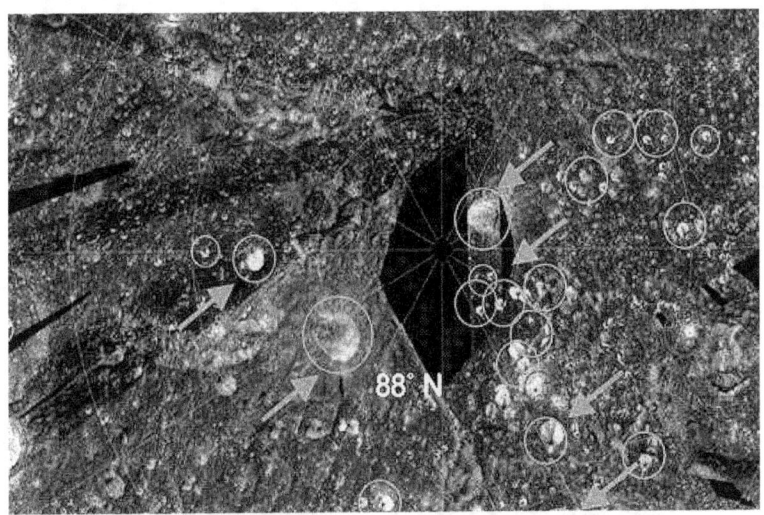

Image: A radar map of the lunar North Pole. Craters circled believed to contain frozen water (NASA)

This is good news for those supporting the alien occupation of the Moon. It would mean humans will occupy the Moon one day and have plenty of water to drink too!

Before NASA ever proved these traces of water, Fred and other scientists speculated about huge reservoirs of water, such as can be seen in the crater Tsiolkovsky on the back side of the Moon (LO IV photos. Picture below, bottom left). Fred said it is loaded with water and is actually a giant lake! "*Notice the large "lake" and the smaller ones nearby.*"

Fred said, "*Clouds can be seen north of the lake. Objects appear to line the North West shoreline.*" Obviously Fred does not know about crater fill-in by lava flows. The crater is very old and appears to be filled in with a flat surface material. He said NASA does not know what the material is. Well, of course, this is Apollo-8, which never landed nor took Mare samples.

He said, "*NASA states that the black mass in this crater is unknown, but that it must originate from the interior of the Moon. It obviously appears to be water, but since this would upset most of the scientific world, the best thing is to wait as long as possible to reveal that*" (Page 109-110).

From a long distance and at the right angle and lighting, it can look like a lake as he said Astronaut Aldrin mentions: "*When I looked at Tsiolkovsky Crater, it reminded me of a mountain lake with a quiet surface and with a small island in the middle"* (page 110).

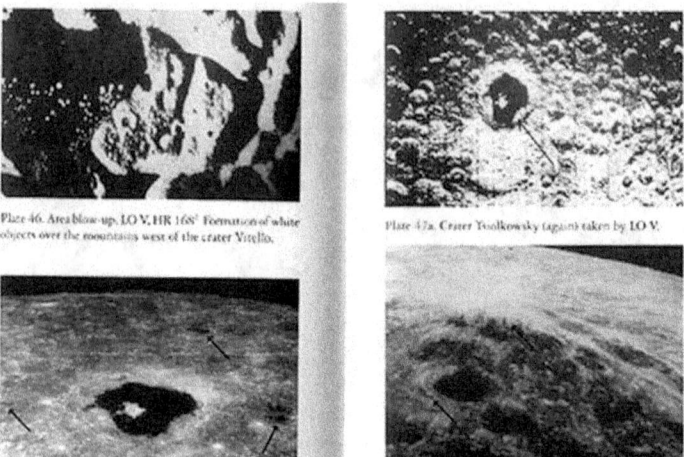

Bottom Left: Apollo-8 photo of the crater. Top Right: Tsiolkovsky Crater taken by Lunar Orbiter IV. Bottom Right: Apollo-8 photo of a crater he calls a "lake" on the back side of the Moon close to the North pole. Bottom Right: More crater lakes.

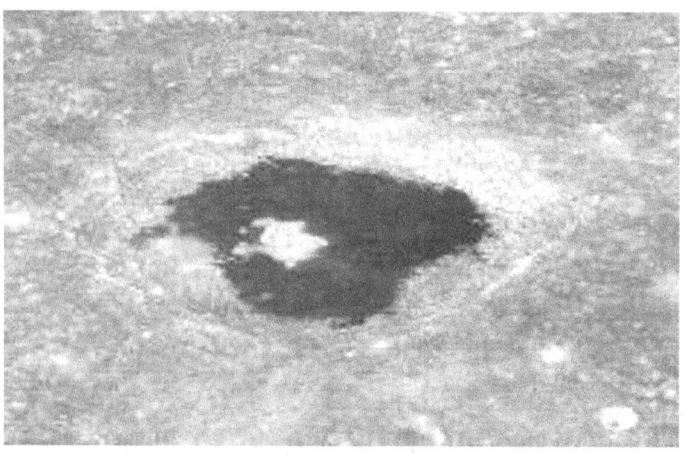

Apollo-8 Photo

14
TSIOLKOWSKY CRATER WATER DROUGHT

Close up of Tsiolkovsky Crater

Fred argues that these are lakes built by aliens for water storage. They left them openly visible for us to see when we were to ever go to the Moon!

Here is a close up of Tsiolkovsky crater. I wonder what those small holes are in the surface of the water. Let's take a closer look? Wow, more small impact craters! There is no water. Looks like a complete drought to me. Maybe we are looking at a dry sea bed?

AS17-139-21302HR

Tsiolkovsky Crater floor littered with small impact craters

Maybe they are floating craters? Holes in the water or alien boats? Or, maybe they are just what they look like, craters in the floor of Tsiolkovsky, peppering the surface like gun shots.

It is amazing how a so-called scientist, Mr. Steckling, avoids close up imaging and enhancements. It's all the same blurry medium shots that are printed out of focus. Anyone can take a photograph of a crater, snapped from a medium or long range distance and make out all kinds of things.

Alternately, when you ZOOM in and see the details, all you see are craters, "regolith", boulders, rocks, dust, fault lines, rilles, cracks, as well as mountains and valleys, and what looks like areas where water or lava once flowed. Other NASA photos show craters that look like lakes: See AS8-12-2209 and AS8-12-2196.

An atmosphere was once on the Moon! Fred said that water was once on the Moon in huge quantities! Maybe so. Even NASA has made such claims. But an atmosphere that still exists? Fred said, NASA is telling lies to throw off the public, because it would scare us! That they want to hide the truth about lunar rivers, lakes and ponds. Such tremendous knowledge would destroy our present thinking.

Fred decides to show us, lakes, rivers and ponds to debunk NASA, who said there are no such lunar features. Yes, there may well be water, ice and other frozen gases, but as to gushing rivers, water falls, oceans and plants? Let's take a trip rather into the hollow earth and search for giant subterranean mushrooms!

The World's Largest Mushroom!

15
MOLEHILLS, MUSHROOMS AND MICROBES

Here is a big long emulsion splatter running across the whole photo. Surely it looks like some great storm crossing over the lunar surface causing the source of lunar water.

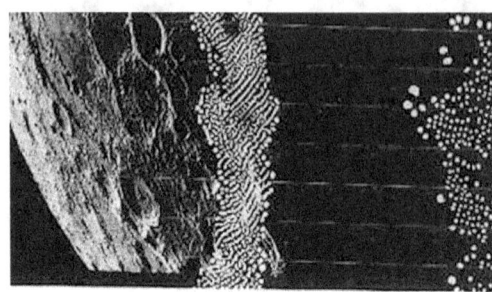

L0-V-65-M Largest Microbial Colony in Outer space

Emulsion splatters, people! Look at the evidence for yourself and stop swallowing exoquack theories for truth. Without investigation and verification you can end up believing anything.

What about space born viruses and big huge bacteria-like "things" floating over the Moon? More emulsion problems? Look at this shot from Lunar Orbiter V-22-M. [See Lunar Orbiter Atlas, plate667].

They sure look like viruses floating around the Moon. If one were to see these in a fanciful alien theory book, without proper reference, he would believe they were. Actually, the internet is loaded with hundreds of web sites claiming all sorts of science fiction theories. The higher resolution the images get, the closer and smaller artifact hunters search and thus, the smaller the artifacts become.

This is no great wonder to the true scientist who knows that as higher resolution photos are obtained the more the blurry out of focus older photos are debunked of having any alien evidence.

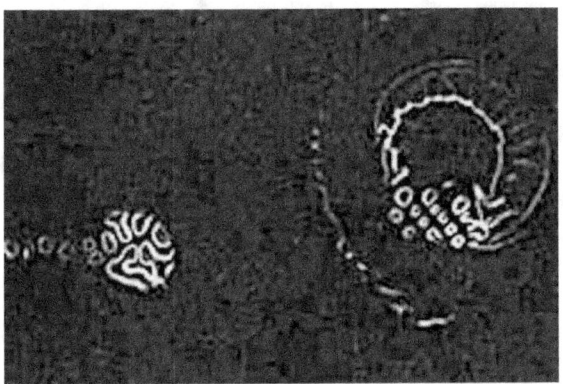

More Multicellular Emulsion Amoebas

**Mysterious Alien Flying "rod" Spy Device?
Be careful what you read and believe now days!**

    This easy to believe syndrome is even more wide spread now days with alien surveillance "rods" flying around the earth at light speed! Ever hear of "rods"? Well, they are weird looking spiral devices sent here by aliens or they are micro-size alien creatures of some kind. No one really knows. Maybe they are photographic glitches?

    In the next section we will deal with more lunar laughables with the studies done by Mr. Swaney in "Objects on the Moon."

SECTION-4

JACK SWANEY'S
"OBJECTS ON THE MOON"
(Fate Magazine, June 1982-84)
Intermediate Idiotic Incidentals of Intergalactic Garbage

1. MARE CRISIUM CONTRAPTION

LEVER STICKING OUT OF THE MOON
Located at 50 degrees E. Longitude, 15 degrees N. Latitude.

Jack Swaney, of Fate Magazine, decided to jump in on the extra-terrestrial artifact bandwagon in a series of published articles. As an amateur astronomer and fan of Fred Steckling and George Leonard, Jack decided to be nimble, and was quick to secure his fame, fate and fortune by finding some more alien junk overlooked by his predecessors. There was plenty of ET debris and manufactured material missed by them to be found, lending further credence to the alien presence lunacy, and leading money to the bank account.

In scanning the moon through telescopic observations, Mr. Swaney was aware of objects and formations so bizarre, as to beg description. For example, he found a 45-mile long rectangular pit or housing on the western shore of the Sea of Crisis. This was something so amazing, obvious and apparent, that one cannot help but ask why previous artifact hunters did not see it. The blindness to such bizarre formations begs the question how they could have so easily overlooked them.

This particular object has a large lever sticking out of it. The rod and shaft, comprising this contraption *"shimmers with a metallic gleam."* If this thing actually is a lever, Swaney said, then "weird connotations come to mind." He asks, *"Who or what could actuate such a gigantic thing and what does it do?"* Jack offers the above drawing (rather than a photograph) as demonstrable of the artifact.

One may ask why Jack did not offer any photos of such a gigantic and most visible object. Is the mechanical contraption found in any NASA photos? Checking the coordinates above we turn to the Lunar Orbiter photos for verification. LO-IV-191-H3 shows a nice shot of the whole Mare Crisium area. Swaney published his drawing upside down, with south pointing up to northeast. If we rotate the frame 191-H3 to fit, we see that the only shoreline object that could fit his description is within the yellow circle.

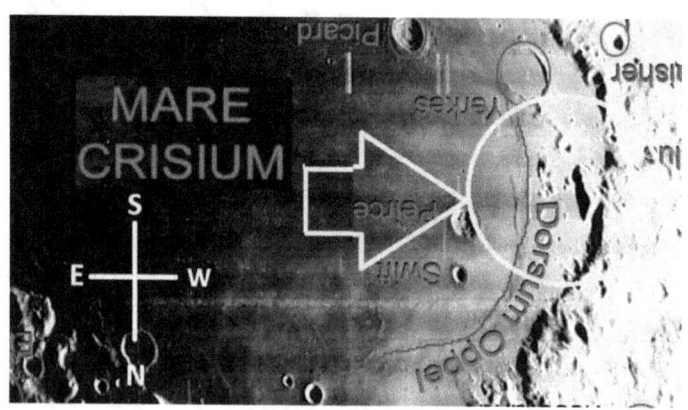

LO-IV-191-H3

Mare Crisium "Lever" object supposedly parked on western shore area.

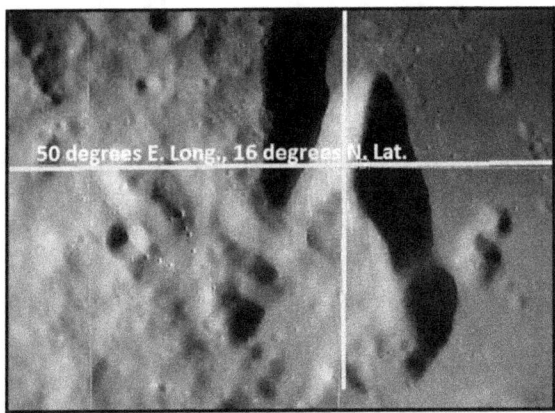

Mare Crisium object - close-up
LROC Act-React Quickmap:
[http://target.lroc.asu.edu/da/qmap.htmlhttp://target.lroc.asu.edu/da/qmap.html]

2. ALIEN CRISIS ON NORTH SHORE

Strange Structures on the Moon in Mare Crisium
(Swaney drawing of Lick Observatory photo 5/7/1938)

Persons who look at the moon through telescopes sometimes see formations that resemble artificial buildings and constructions unlike anything known on earth, said Swaney. One area he points to is the north shore of Mare Crisium, where a small round spot sits just above right center on the moon's face. He said there are a number of them sitting on the southern and eastern shores, and two on the northern sector. South of Crater Macrobius *"stands a quaint round structure with a peculiar slant, measuring 25 to 30 miles in diameter."* He describes the object as *"a cannon"* that *"protrudes from its topside at the back, while half a dozen openings ring the front."* Below Crater Cleomedes about 100 miles east is another anomalous formation 18-miles-long, rectangular in shape like a building that has six silver-white spheres sitting on its roof. Again he draws a picture rather than give us a photo.

Jack proposes that the government has covered up the truth of these objects by suppressing older observatory photos predating 1950 and has actually confiscated them. Persons searching NASA photos for these structures will be inevitably disappointment, if they are expecting to find anything. The truth is that anyone looking for Mr. Jack's lunar candlesticks and sand castles will be disappointed no matter what photos they dig up or what telescope they look through. Yet, at least one photo still exists from Lick Observatory, from which Mr. Swaney drew his illustration above.

A comparison of NASA photos will solve the crisis of the missing mystery photos and the vanished alien constructions. They are not there now. They were not there before and they will not be there any time in the future – they are fictions of the imagination of the writer of the Fate magazine articles.

The following photos show the area between crater Macrobius and the smaller crater named Peirce. The objects drawn between Macrobius and Peirce are interpretations of the shadows and sun highlights playing off the surface features of the moon. It seems the "cannon" with window openings has moved on to another location. As for the 18 mile size building 100 miles south and east of Cleomedes, it too must have been relocated somewhere else. The red square in the following photo outlines the closest thing that looks like what he is describing – a lunar highland ejecta area between craters.

LROC Act-React Quickmap, northwest Mare Crisium, an area between Craters Macrobius and Peirce.

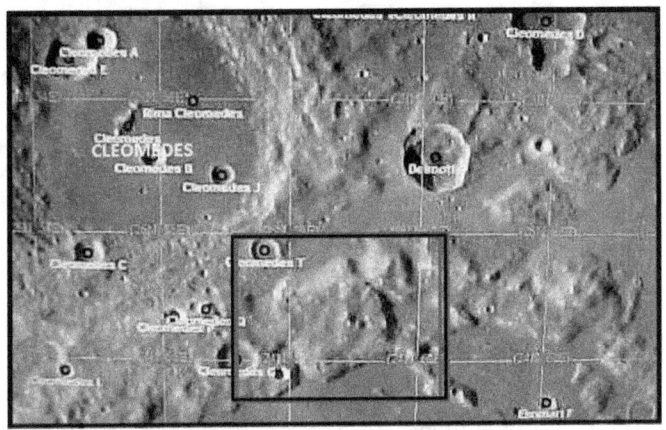

**Crater Cleomedes area showing no building.
LROC Act-React Quickmap**

3. THE GIGANTIC BLOCK IN ENDYMION

WRECKED OR DISCARDED artifact on the eastern rim of Crater Endymion at 54 degrees N. Lat., and 60 degrees E. Long.

In the northeast quadrant of the moon lies a strange rectangular junk heap well over 50 miles in length. It's a double-decked ocean liner size square block, tilted on its side, giving the impression of a beached whale washed up on shore. In this case, it is a mechanical crater critter resting on its side, moon bathing. It has four brilliant white domes, two of which seem to be up on poles and a circular superstructure on the upper deck.

The above monumental heap pile and mechanical mishap depicts what Swaney said is evidence to as long-concealed truth regarding our solar system's recent history, things our scientists refuse to talk about, which happened some 12,000 years ago. Swaney said, this was coincidentally contemporary with the disappearance of Atlantis. Jack claims the evidence is surfacing in support of ancient alien technologies and historians will have to give an account – I.e. rewrite the books, restructure religion to accommodate aliens and redirect humanity away from godly truth. It is easy to deny the existence of this technology said Swaney. Yet, if considered, it is not easy to pin point its demise.

One explanation, given by Fate Magazine, is ancient Greek mythology. The Greeks tell of a prehistoric period called the "golden Age," when the gods ruled the heavens and earth. The myths tell us that a colony of renegades known as the Titans invaded this solar system with intent to conquer. The ancient Atlanteans fought against them to no avail. The Titans used nuclear weapons and defeated the Atlanteans. They blasted away at the smaller planets and blew up Mars, destroyed the planet between Mars and Jupiter, and disrupted some of the gas planets like Uranus, capsizing it the way we see it today.

Based on the false assumption and pseudo-science of the atomic bomb explanation of crater development, the above (exaggerated) interpretation would seem reasonable. Nevertheless, having some reasonable shadow of a doubt, there must be some other reasonable scientific explanation.

The science of geology is a mystery to most people, who would not know a conch shell horn from a piccolo. Geology explains exactly what happened. The above mechanical mirage is a byproduct of light and shadow. This drawing was made from observatory photos or from direct telescopic observation, because none of the NASA photos reveal such nonsense. The article does not reference any lunar photo number. Nevertheless, if the object was there in the past, it is not there now.

The LROC Act-React Quickmap shows the location of what may be mistaken for the above drawing. Improper lighting and shadow play can easily form artificial looking constructions. The observer can imagine giant size monstrosities from this trick of light and shadow, and if careful, the sizes can range into the hundreds of miles.

The above square shaped colossal size contraption has to be hundreds of miles long and exceeds the limits of sane engineering, even for alien cultures. It seems the further away the alien presence artifact hunter is the larger the artifact. Conversely, the closer the snoop finds himself, the smaller the alien artifact. The only common denominator would be the blurriness in hiding the true nature of the object.

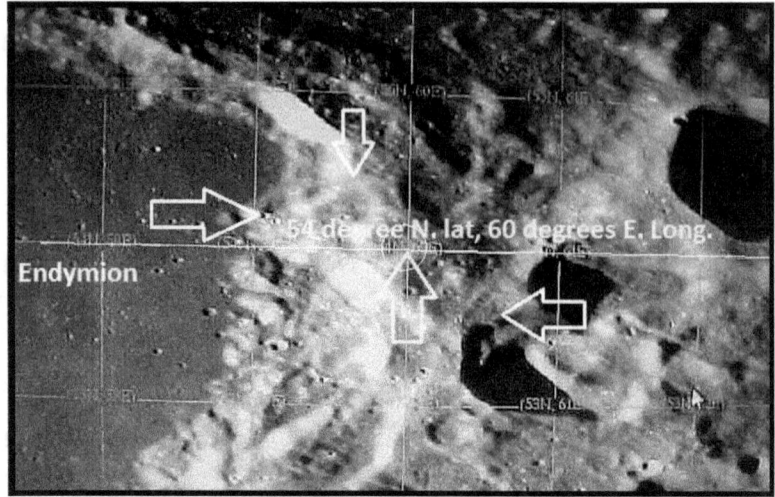

Eastern rim of Crater Endymion

4. ROBOTIC BUNNY IN CRATER CLAVIUS

Kewpie Doll lying on southeast wall of Crater Clavius
(Location: 60 degrees S. Lat., 8 degrees W. Long.)

No, this is not a funny looking robotic gadget or some other alien claptrap. The weird shape looks like what Swaney and Felix Bach said is a *"discarded kewpie doll."* Wow! Giant Chatty Cathy's for giant aliens! The illustrious reporters declare they have seen this surrealistic toddler sitting close to the South Pole on the wall of Crater Clavius, cocked at an angle as if knocked off its base. Some kewpie dolls have support bases. This one has a base attached to its bottom. This bizarre doll-baby has a metallic sheen like dull aluminum and is visible in some observatory photos of the moon.

Nevertheless, one should not hold their breath in finding the kewpie kwickly. Most of the photos have been touched up, repainted or confiscated. Do the NASA photos expose this E.T. infant in any of the millions of pictures taken of the moon? LO-IV-118-H3 and 131-H1 do not show much supporting evidence. Neither does frame 124-H1 and neither do the 10,000 plus other NASA photos.

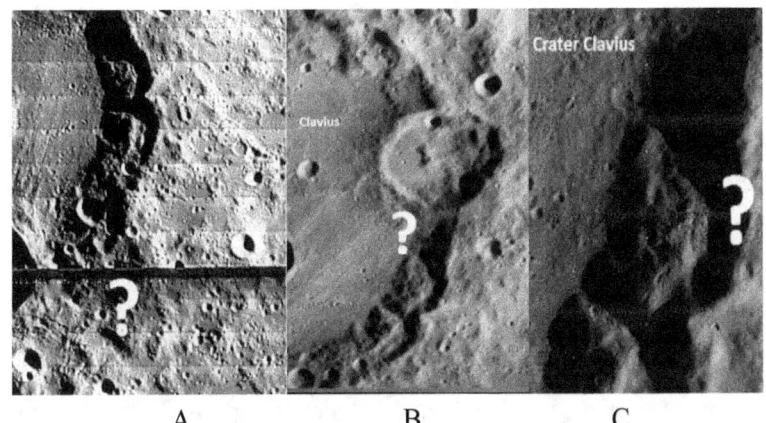

A   B   C

LO-IV-118-H3   b. LO-IV-124-H1   c. LO-IV-131-H1
Southeast corner rim of Crater Clavius

5. THE RASCALLY RABBIT IN THE EAST

Rabbit Ear machine east of Sea of Nectaris
(Located at 25 degrees S. Lat, 45 degrees E. Long.)

Hi Ho the Dario, a hunting we will go and to the east quadrant of the moon too! We are hunting "rabbits" in the east or one-half of an ocean liner's prop. A rather bizarre object is supposedly in the moon's east quadrant, a little east of Mare Nectaris close to Crater Theophilis. Drawn from telescopic observations, the mechanical rabbit appears a dazzling white, almost fluorescent in brilliance.

The above photographic hair hunters insist that when it is near the shadow line, it is far and away (by far) the most immediately visible feature on the lunar surface. They speculate that, just like a real rabbit, the ears of this mountain size Lagomorpha-leporidae is intended to move, because they are mounted on a vertical shaft, which rises above the ground.

This above interpretation is a "hair" bit above ridiculous, as photo references do not show any such hair-brain rabbit device. Actually, it is by far the most elusive rabbit on the moon. None of the NASA photographs catch this bunny tale and reveal only highlighted lunar craters, regolith and shadows. This is another wild hair story and goose chase among the craters as the following photos suggests. Nowhere do we find this White Rabbit.

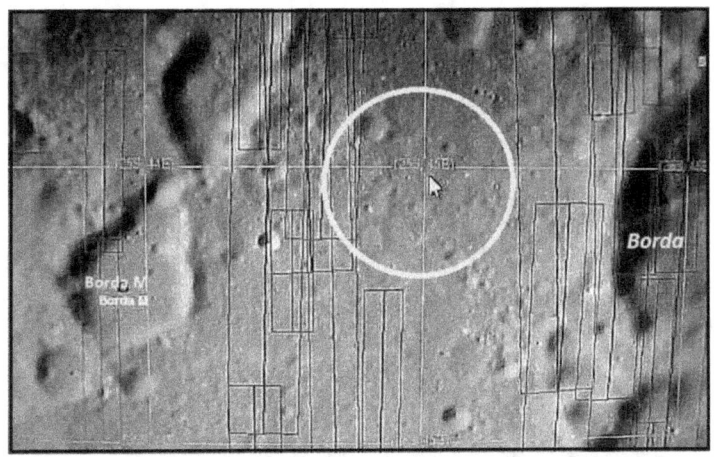

LROC Act-React Quickmap of area southeast of crater Theophilis. Location of the elusive white rabbit.

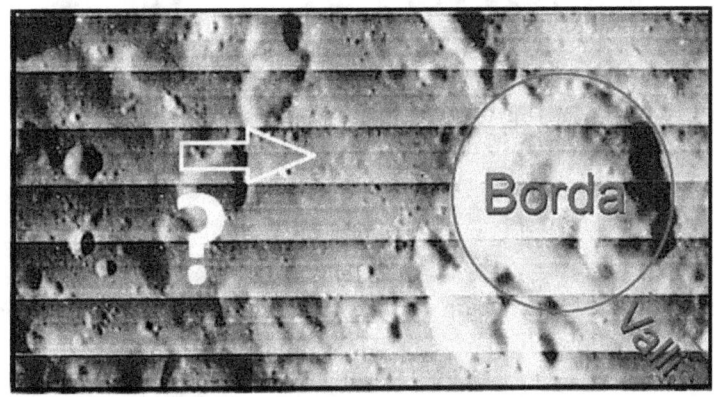

LO-IV-065-H1
Rabbit location E. and S. of Mare Nectar and west of Crater Borda.

6. MONUMENTAL MOUNTAINS OF MOON METAL

Piles of malformed metal heaped up on the south tip of
the Caucasus Mountains, along the edge of the Sea of Serenity.
(Location: 35 degrees N. Lat., 8 degrees E. Long)

Our friends at Fate are on another wild salvage hunt and this time for scrap metal. In the moon's northern hemisphere (northeast quadrant) resides a junk pile location littered with twisted metal plates, broken constructions and bizarre chunks of machinery, stretching for 200 miles within a lunar formation named the "Caucasus Mountains." It is supposedly the amateur astronomer's favorite telescopic view for its dazzling piles of shinny geometric shaped chunks of rubbish. It is a real junk pile hill for possible ULO's ("unidentified lunar objects").

The Caucasus Mountains are in the Sea of Serenity, which lies southwest of Crater Calippus. Rukl's Atlas in map 13, p. 53 demonstrates a good shot of this location. Other good shots of "Junk Pile Hill" are in NASA photos LO-IV-103-H1, H2 and 098-H1. It is up to the reader to decide the location of this pile of fantastic garbage. The area is steeped with what looks like alien junk and just about any lunar-graphical location will expose the stuff.

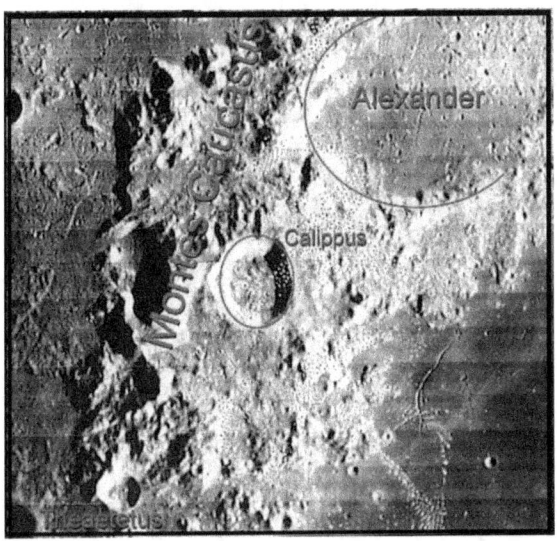

LO-IV-103-H2

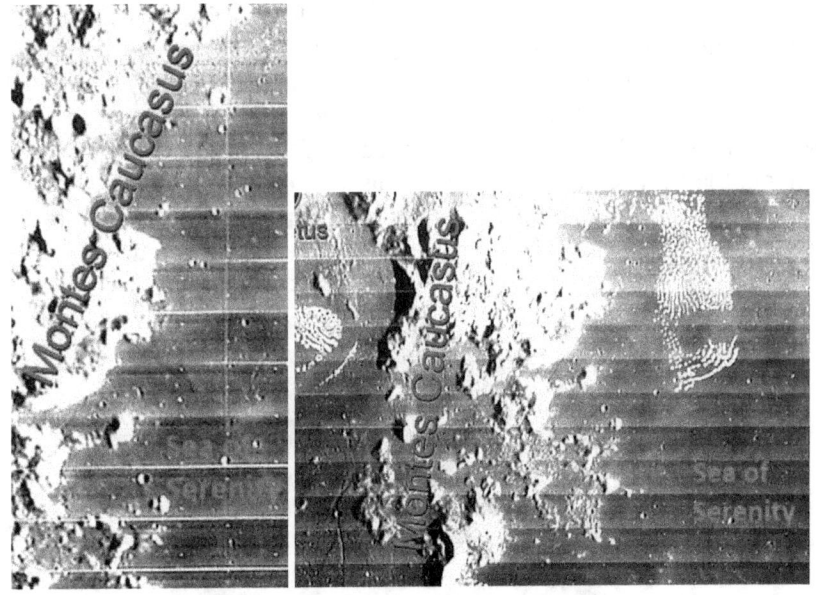

LO-IV-098-H1   **LO-IV-103-H1**
(Lunar Orbiter Photographic Atlas of the Moon)

7. CRANES AND BOOMS OF JULIUS CAESAR

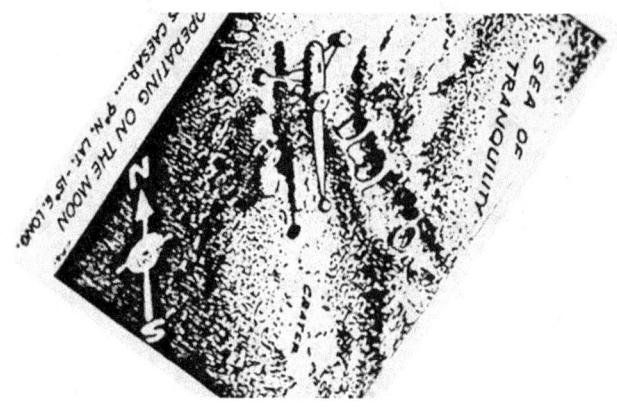

A 25-mile long crane or boom is proportionally unthinkable and it is unthinkable that people should believe such hog wash. The above is a sketch of some machine made in April 1982 by the Fate observers, when the moon was only six days old. The writer pontificates that it is located at the north rim of the Crater Julius Caesar and easily viewed at 9 degrees N. Lat., -15 degrees E. Longitude, on the western edge of Mare Tranquillitatis. The ball-tipped arm is usually orientated north to south, but on occasion reorients itself east to west. In other words, it moves! Maybe it is light sensitive and changes shape and location according to the Sun?

LO-IV-090-H2
Crater Julius Caesar
No rotating weather vane. Just a mountain range with dark shadows.

Previously, in 1980, the astronomical weather vane rotated 180 degrees, the ball-tip riding above the semicircle of structures in the foreground. The blab continues with braces that apparently steady the front end of the boom, which are ahead of the pivot point. This suggests they have to be unconnected before it makes the 180-degree swing and there is no one up there to do that. Another gigantic moving "thing" lies 75 miles west close to Crater Boscovitch.

SECTION-5
MOON MARS MONUMENTS MADNESS

A POLEMIC DISCUSSION OF RICHARD HOAGLAND'S "THE MONUMENTS OF MARS: A CITY ON THE EDGE OF WHERE EVER"

Mr. Richard Hoagland

The most prevalent and modern equivalent of George Leonard and Fred Steckling is Mr Richard C. Hoagland, self-educated [1a] scientist and famed author of "The Monuments of Mars: A City of the Edge of Forever," and "The Enterprise Mission" web site. Mr Hoagland is a former space science museum curator; a former NASA consultant and during the historic Apollo Missions to the Moon, was science advisor to Walter Cronkite and CBS News. At present he hides behind a web site with no direct phone line, runs the country giving talks and lectures and sometimes visits NASA and other government agencies promoting the godless alien presence mythos.

Following the earlier tradition of Lunar surface alien trivia, Mr Hoagland thinks he sees extremely complex mathematical and hyper dimensional geometry in such things as the colossal *Face* on Mars and surrounding rock piles of Cydonia. He also sees this strange type of alien stuff scattered about the lunar surface of our moon. For instance, Richard asks a rhetorical question, surly not a real question (for he really believes in alien artifacts), that *"If a complex, hyper dimensional, spacefaring culture built, and then abandoned, the monumental ruins at Cydonia,* [then] *what other artifacts, on what other worlds, for what hyper dimensional purposes* [and he does not really know], *might also have been left across the solar system, including on earth's Moon?"*

The Mars Face was [at that time] a sure thing, but the lunar evidences were a real nut to crack. Richard was reluctant to dive into the Sea of Tranquility and disturb all the previous questions of crack-pots, for he was still without [and still without to this day] *"some kind of indication as to where to search, to even begin to pour through the literally millions of NASA close-up images taken of the Moon."* To him it was all but hopeless, but to others it can only take maybe a dozen photos to prove or disprove the alien artifact urban legend.

No matter, Richard persisted and *"Only after beginning a* [caught, choke, strain to not laugh] *serious investigation into a possible hyper dimensional influence between Earth and Moon, did we also inevitably begin consideration of a crucial corollary: the possibility for ancient ruins on the Moon, left by those who long ago left us this new physics."* The problem he said though, was where to find them. Strain as he did and so easy as it came to be, he then began to find more alien artifacts than George and Fred could have ever found. Almost every photographic scratch, glitch, dust particle and especially dangling crystal (development scratch) on the photos became an alien artifact.

Unfortunately for Richard, as it was for his predecessors – thanks to Richard's expose', these artifacts are as well fictional inventions created from hyper dimensional, delusional thinking – not much different than the hyper-two-dimensional thinking of the two previous researchers.

Of course, it took Richard some time to overcome the lack of "real" evidence, supposedly found by previous researchers: i.e. George Leonard and Fred Steckling, whom he purposely criticizes and debunks for proposing that emulsion development processes were alien artifacts, to come to grips with and supposedly prove that tons of photographic pixel distortions are rather the real alien artifacts! One thing Richard did maintain in common with his debunked predecessors, in establishing an alien presence, was the same old light and shadow play on lunar surface features that deceived Leonard, Steckling and many others.

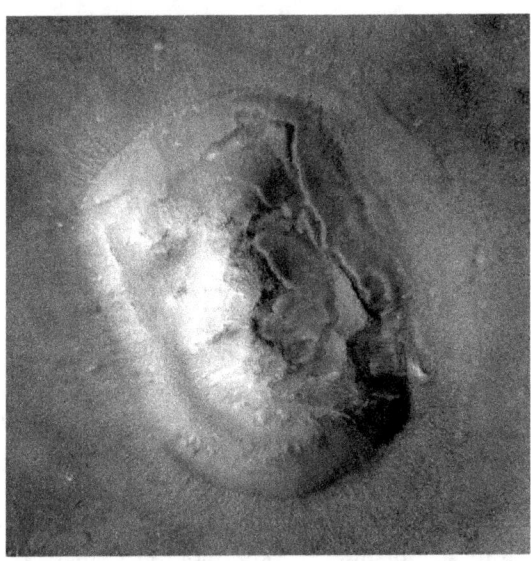

LROC Photo of the famous hyped up "Face" on Mars

The Mars Face is a prime example of this deception. What was once a real face and the foundation to a whole theory has recently been proved with higher resolution photos to be a giant mountain size pile of Martian rocks. Thus, it is this writers opinion that everything that follows the Mars Face as evidence for aliens is just more advanced piles of Martian regolithic rhetoric.

The Mars Face (illusion) borders the plains of Acidalia Planitia and the Arabia Terra highlands. [1] The area includes the Mars regions: "*Cydonia Mensae*", an area of flat-topped mesa-like features, "*Cydonia Colles,*" a region of small hills or knobs and "*Cydonia Labyrinthus,*" a complex of intersecting valleys. [2]

This is where Richard goes monumentally mad. His in-depth studies of lunar and Martian alien artifacts overshadows the combined researches done in previous decades. In fact, this is exactly what he has done, combined all previous studies (minus the "water-droplets" and "pull-apart marks") adding further a hodgepodge of much new material – mythologies, fictions, imaginative inventions, typologies and abstract concepts, all gift wrapped in super scientific mumbo-jumbo. The consequences of this man's work has led to a global wide pandemic of alien artifact finds by what we may call Hoaglandites.

According to Hoagland and his cult followers the Moon is littered more now than when Leonard bent the first lights of Selenography studies. After bleeding the Man in the Moon dry, and to escape boredom, many E.T. hunters jump this new train to other planets in search of the fabulous.

Mars has been found to house new alien landscapes for mad monument hunters. The scientific realm, especially the space programs of our nations are cluttered and mingled with this alien pandemonium of non-fact and nonsense. I suppose that the portion that does not believe is more than happy with all the publicity and has rather turned the other cheek caring less about debunking it. What a loss of publicity and funding! So, why bother?

Mr Hoagland's supposed revolutionary concepts, stemming from the Cydonia region, suggests an astrophysical origin and role of our Moon, and the possibility of finding alien artifacts. He sees remnants of an alien culture that once thrived in this area and supposes that Mars was once a global cosmopolis of buildings, pyramids, super highways and advance glass tube tunnel systems. He postulates that if a hyper-dimensional space-faring culture built the Cydonia Complex, then we all must ask, *"What other artifacts... might also have been left across the solar system, including the earth's Moon?"* [3] We will find out the answer in the remainder of this compendium.

To answer the mute question of where these Martians went to next, after cratering Mars beyond livable conditions, we must turn to the Moon, where we see they also cratered the lunar surface and disappeared. We dare not ask where in Hell they went after the Moon. But that might be a good place to start looking! With a full knowledge of Earth history and a total lack of earthly alien artifacts to suggest that Earth is where they went to next, Richard turns our attention as well as our stomachs (without much mention of Earth) to the next best location – the Moon.

Of course, the moon! Where else can one play with people's imaginations, twist landscapes to beef-up the alien presence cult and create a super income for himself as well as advance public interest in NASA? As far as NASA and Uncle Sam are concerned, they have their Mickey Mouse now, thanks to Hoagland. Who would ever dream of flushing Mickey down the toilet in favor of raw dead dry science?

Richard has only three reasons to choose the Moon. Each is perfectly strong enough to deceive the ignorant masses. First, the Moon is preferable, because the Earth has no super technology, "super" physical enough to prove an alien presence in early human history. What it has is only suggestive and not at all tangible. There are no huge colossal crystal glass domed cities, no mile high alien monuments or mysteriously advanced non-human scientific devices.

All that we do have are vague references to a high technology in our ancient past that any culture could develop within four to five thousand years of scientific evolution. Hence, ancient advanced cutting marks, sonic drilling and other weird and unexplainable artifacts are only so many evidences for advanced human engineering.

There is nothing on earth that cannot be explained by human development if one is patient enough to look and avoid jumping to alien conclusions – give or take a few mutilated cows or destroyed corn crops – and why only cows and corn crops, and not chickens and cabbage we will leave for another volume.

Nevertheless, when it comes to "evidences" found 250,000 miles away, shot with not so fine photographic equipment, anything can be wizard into proving an alien presence. Shadows, highlights, warped surface features and other natural lunar-physical anomalies are easily contorted into alien manufactures.

Now, unknown to the common alien Oopart hunter, just the opposite can be proved, which is even truer, when the same materials are exposed to closer inspection. Domes, machines, towers, pillars, obelisks are easily seen to be mounds of dirt, rocks, boulders and other naturally developed selenological features.

Secondly, up until recently, the lunar surface had been studied in more detail than the Martian surface. Since the earth was void of alien Ooparts, and Mars was still obscured by low resolution photography, the Moon was the only best choice. It was far away, mysterious and yet close enough and available to study. Since the Viking missions, the Moon though was abandoned in favor of the more romantic new world of Mars.

Coincidentally, as Martian photography advanced to what it is today, and as lunar photography developed even further in resolution, the direction of people's consciousness moved away from the Moon, and for good reasons too. Lunar photography is even higher in resolution than the Martian photography, coming from the Lunar Reconnaissance Orbiter, and lays waste to any attempt to discover any alien artifacts, as we have seen.

We shall see in the following studies that the advancement in photographic resolution of lunar surface features has exposed all this alien artifact trash as pure nonsense, and all the theoretical nonsense as pure trash. However one prefers to mock it matters not, it is now being proved with Mars. As the photography has gotten better, so also has the exposure of Martian artifact nonsense.

A grand example that is unfortunate for Mr Hoagland (and he should have never pressed the issue of demanding finer detailed photos of Cydonia) is the Face on Mars. As we have mentioned above in the most recent photographic missions, the Face is proved to be what real science predicted – a giant pile of Mars rocks.

We shall not choke these pages full with all the double-talk of Mr Hoagland and his followers, as they try to explain the blatant evidence against the Martian face. The Internet is full of Richard's expos-fact-o sophistry. In fact it has become a sewer of debates, most favoring the Hoaglandites, for people prefer the fabulous over the facts.

Getting back to the original point: The Moon in past times was closer and offered more detail, yet still vague in resolution to really decide for sure the reality of alien artifacts. Hence, we now understand Richard's quick distracting disconnection from his predecessors.

These studies offered (at that time) the necessary *"conclusive evidence of contemporary extraterrestrial bases on the Moon."* [See footnote 3] This was the one and only and last chance to seriously investigate and embed the alien presence syndrome into people's minds, and thus prove Hoagland's so-called *"hyper-dimensional influence between the Earth and the Moon,"* and of course Mars.

The only thing missing were the facts or artifacts and exactly *"where to find them."* [See again, footnote 3] Since they were NOT good enough, though they were good enough as a starting point, Richard decided to replace the old theories with his new physics.

With such a vast surface area to look over, Richard decided to pick up where the others left off. The most convenient *"jumping off spot"* were the previous studies done by such alien Oopart hunters as Fred Steckling [4] and George Leonard. [5] After some lengthy digging through all the *"non-stop comedy of errors, misidentified photographic defects and the general mishmash of malillustrated misinformation"* [6] of these forerunners of alien artifacts, Richard decided that it was his turn to advance the non-stop comedy of errors, misidentified photographic defects, pixel distortions and the general mishmash of malillustrated misinformation with new examples, not only from the Moon but from Mars.

Though he *"looked hard... but could see nothing"* of what they were claiming, it was not long before he looked hard and began to see something. To him, just as it was for his predecessors, something was very *obvious*, yet easily missed, if one did not have a hyper-dimensional mathematical eye to see.

Of course, anyone can develop the hyper-dimensional arithmetic eye, after days of staring at blurry photographs of dead dry boring surface features. It is not much different than gawking at Earl Grey leaves and prattling fancies about one's health and welfare, after noon time tea and crumpets.

To save the reader their eye-sight and not belabor the subject beyond sensibility, we shall reduce the Martian and lunar piles of Hoagland hog-wash to a short study, picking only the most obvious and favored, as well as rediculous examples of the new evidences.

1
UKERT CRATER BASEBALL DIAMOND

AS10-32-4819

Marvelously, after analyzing thousands of blurry, dead scientifically dry photographic lunar features, it was not long before Richard's eyes fell upon a most interesting formation immediately adjacent to and somewhat southwest of Ukert Crater. It was NASA Apollo-10 photo AS10-32-4819 that attracted the attention of Richard and it was the best choice and finest starting point for a most obvious and potential lunar alien artifact that could begin proving the alien presence on the Moon. What he found was an almost perfect square baseball diamond shaped looking plot of dirt.

From this default in clear vision followed the delusions found in photo frame number 4822. Just as was conjured from the Cydonia ruins with the Face on Mars, Ukert showed possible evidence for a similar type of "hyper-dimensional *grid*" and thus potential proof of a "hyper-dimensional *alien biological and cultural connection.*"

The Ukert Crater Baseball Diamond Field

This was a find as grandiose and colossal as the Giza pyramid of Egypt, the Temple of Baalbek, and the ruins of Teotihuacan. It was equal to the famous Seven Wonders of the Ancient World! This was an ancient alien city of the edge of forever or for sure, the edge of Ukert. It appeared long ago abandoned and forgotten. Yet, something was there! What was it that caught the attention of Richard's eyes? It was a little *"geometrically marked square"* next to a big crater.

Baseball Field from Top View

With heart in mouth, an apple in his throat and a pen in his hand, Rich began to analyze this remarkable compact, bluntly obvious geometric arrangement that was leaping off the page at him, and him alone. For, after some further attention by this writer of the same photo, and after what seemed to be an eternity of looking, nothing could be seen to verify such claims of alien biological cultural connections but a square plot of dirt created by possible crater ejecta materials.

Nevertheless, Richard persisted and with ruler in hand, and some scientific thought, or what he thought was scientific thought, he applied to this tiny geographical region of highly fascinating, but highly anomalous organized dirt some super mathematical calculations. Richard points his own Leonardean magic wand at absolutely nothing and instantly transformed this regolithic nothing into a megalithic something. A close examination from different angles from all the relevant photos of Ukert revealed to Richard a magnificent alien construction, whereas the same study repeated by this writer revealed absolutely nothing. See As10-32-4813 through 4822. [8]

Further study with a ruler in hand, a calculator, a micrometer, a spectra-meter, an oscilloscope, an old but usable electro diagnostic violet ray wand, some Flemish, English and French ell, fathom, feet and cubit tape measures, this writer found some odd and offbeat "moot distances," yet coincidentally similar to the smoots (364.4 smooths) the length of the Harvard Bridge.

Yet, all attempts to translate this anomaly into varied other important measurements – megalithic yards, bloits, pyramid and British inches, sheppey, a few Mickey's multiplied by the factor of 1000, furmans, and a few attempts of Potrzebie, cowznofski, vreeble and some hoo and hah – failed to reveal any further so-called Hoagland super hyper-dimensional geometrical equations. There was nothing whatsoever akin to the supposed geometry of Cydonia. There was no stroke of alien Oopart luck revealing Lego type constructed buildings, no electromagnetic hovercraft freeways, no mechanical stairs, steps, walkways, glass domes or used abandoned UFOs. There was only a crater and a funny little (somewhat) square baseball field diamond shaped plot of crater ejecta.

2
RETURN OF THE UKERTIANS

With some extra careful observations, beyond the abilities of Richard, one can notice other supposed obvious and blatant evidences of alien activity or what appears to resemble alien "Ooparts," "Oops," and "Oops, it's busted and left behind" equipment. Glaring at Lunar Orbiter-III photos for hours on end, one cannot help but see, even when hallucinating, what Hoagland missed by not looking hard enough. The aliens apparently never left or they came back and are still using the area. There are five or six small white orbital shaped UFOs floating about the crater!

Interesting little blobs. Maybe the aliens are back observing or salvaging their old properties? Surly with the help of their little fleet of "water-droplets" and "pull-apart marks" floating in zero gravity emulsion development fluids, and at 1/6th gravity (located to the upper right middle in the crater) hovering over Rime Hyginous, the aliens will have no problems rebuilding the city – or at least a baseball team reunion in the diamond square. What of the huge squadron of UFOs to the East?

We are not sure, said people like Hoagland, so they might well be the reconnaissance missions of smaller scout ships hovering south of the larger assembly of the crater blob crafts, and directly north the group of nine crafts running their mission over Ukert City area. Richard! You missed this. These are not emulsion splatters. Because, if you were to snap another picture of the area today, we'll bet you all the tea in China that they are not there anymore – they have moved on to another location – actually, the Lunar Orbiter Photographic Atlas of the Moon is replete with examples of these "things" floating all over the Moon. They seem to be there one minute and then not the next: Now you see them; now you don't.

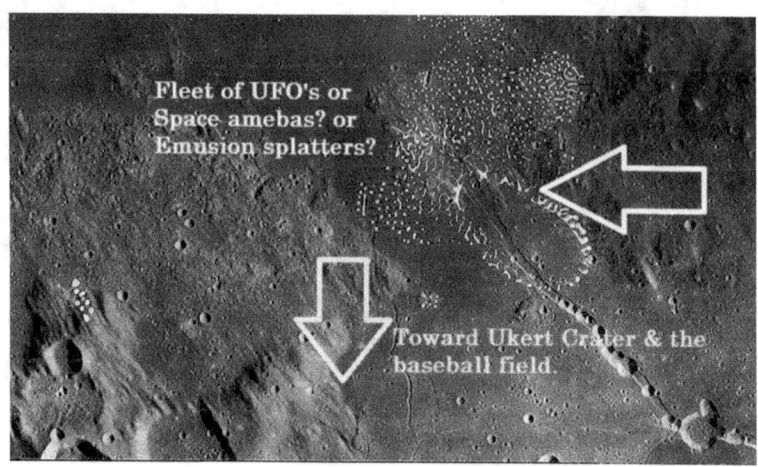

Moreover, if you look at the most recent photos of the same area taken by the Lunar Orbiter Reconnaissance Mission, they are not there either.

Poor Richard's Almanac does not stop here. Richey moves on in his fabulous researches for E.T's and artifacts way northward from the Square to what he calls *"the best yet to come."* Oh, I can't wait!

From the square or Hoagland's way-up-stairs "L-7" to the upper right *"lies a diagonal array of brilliantly reflected hills now suspected... to be the shattered ruins of several ancient, environmentally enclosed 'acrologies' (city-sized architectural ecologies) on the Moon."* Beyond arcology row *"lies an equally remarkable 'rectilinear ground pattern' eerily reminiscent of an array of city-sized blocks... (And) anomalous vertical 'reflections' and curiously translucent 'panels' scattered throughout the area,"* [8] with *"overall, striking 'grid-like' 3-D rectilinear geometry... anomalous curved 'pipe-like' structures... (And) the elevated 'glass-like, geometric lunar construction' ... Sinus Medii dome, (of) 'wavy shower-glass veiling'... 'Of irregular vertical lineation's' (of) 'optically reflective lenses' dangling and hovering over the city area."* After reading and prepping ourselves with Gulliver's Travels and dangerously straining our eyes, we could very well see cross-eyed, *"a shattered crystalline city on the Moon."*

With all this hyper-dimensional flap and blab, and the potential evidence of alien UFO scouting activities of the little white blobs, the Ukertians may be returning to claim their old land grants. But, little white blobs and floating orbs are not the only things dangling around the skies of Ukert.

3
DANGLING CRYSTALS OF UKERT CITY

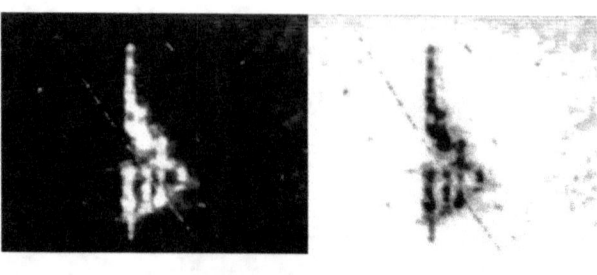

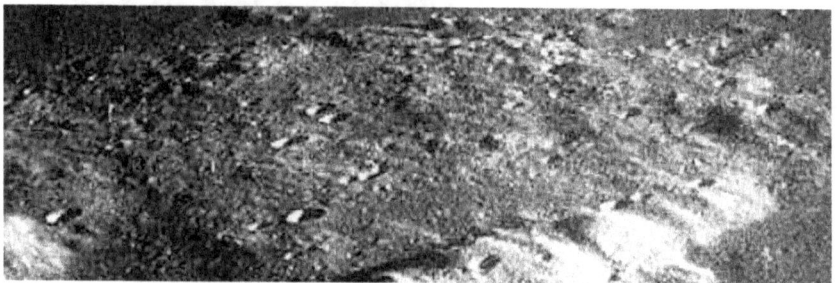

Ukert City, the Lunar Los Angeles

In the Crystal City of Ukert (and it does not matter where one looks in the Lunar Orbiter photo III-85M) one can find what appears to be "crystalline structures" supposedly hanging above the lunar surface, within a *"geometrical grid-like"* system of decayed glass dome support beams and lattices. Now take a deep breath! Our fiction writer said, they all seem to be irregular shaped and unnatural to normal or natural lunar geological structures. Zooming in deeper, both into Mr Hoagland's imagination and into the finer details of the photographic emulsion of frame III-85M, we find what looks like these crystals of wire-like, hair-like bright reflective shapes sticking out like a sore eye hovering over truly square shaped double craters. Here are a few enlargements of the photo and some drawings extrapolated from these high resolution enhancements of the frame, along with some other newly exposed examples demonstrating E.T presence.

When looked at properly, as development processes, and explained as such, it becomes evident that they are not above the Lunar surface, but actually scattered along the plain of the surface of the photo itself. Without taking the hair-like object in context of the whole surface of the frame, it is an optical illusion and deceives one into thinking it is on the Moon.

When taken in context of the frame surface, and when hundreds of other examples are located, it becomes apparent that it is one of many photo development Ooparts or artifacts, even possibly dust on the camera lens, or some other intrusive element. Compare these drawings with the following visual optical illusions.

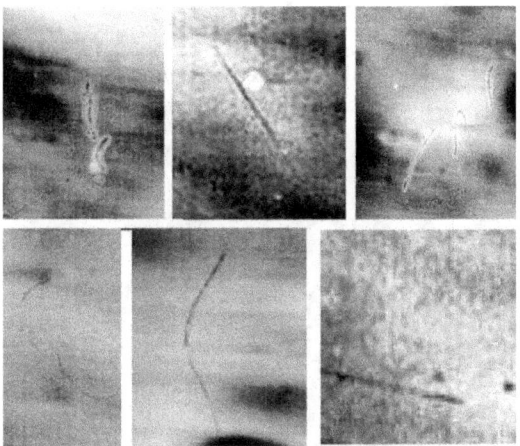

Hairs, lines, scratches and dangling crystals

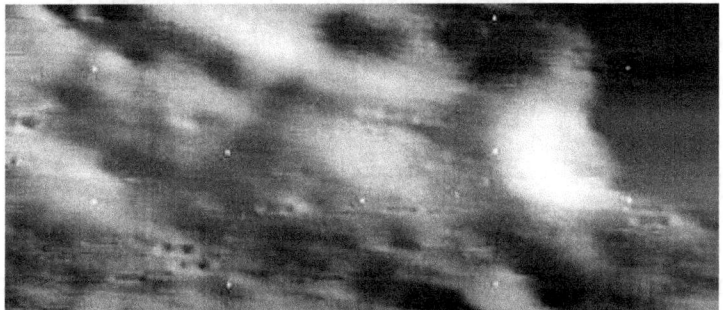

DOUBLE CRATERS

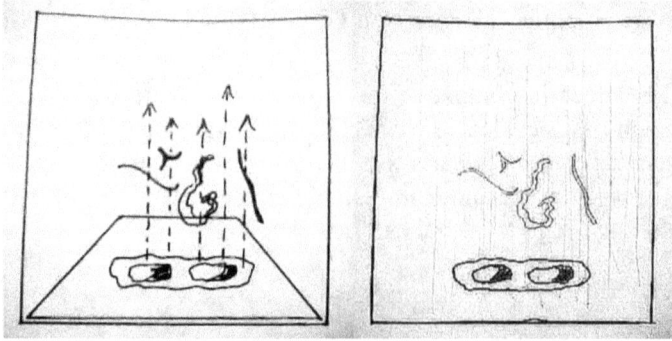

The first drawing demonstrates the way Hoagland wants us to see the artifacts. The second drawing shows the way we should look at them.

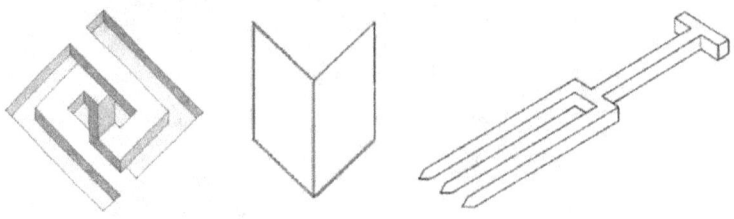

The best example of this eye trick is the middle drawing of the above collection of deceptive constructs. Ask yourself: Is the book open to a page inside the book or are you looking at the cover and spine on the outside? Now turn back and look at the so called crystal slivers and alien artifacts. Is this really a chunk of crystal glass from some ancient domed structure or some other type of alien building material "suspended above" the lunar surface?

Are they really dangling crystals of glass or plastic hanging from geometrically shaped structural support beams? Yes, said Richard Hoagland. But, this is only a hair-brain comic book idea postulated (or hatched) in his overly imaginative mind. No, they are no more real artifacts than the ones pointed out by George and Fred. And they are not dangling in mid-air over the moon. This is an illusion

From Richard's descriptions it would seem more than an idea and actually an alien artifact. Richard is not spinning a yarn or blowing fantastic glass bubbles. He, maybe unlike the first two yarn spinners, actually believes what he is describing. As George Leonard made dirt digging machines out of boulders and as Fred Steckling made ponds and lakes out of emulsion splatters and crater floors, so Mr Hoagland overly imagines and hyper exaggerates misidentified emulsion errors, pixel distortions making Ukert monuments out of lunar molehills.

This is not just some alien hairy crystal hanging above the Moon, but actually a "hair" or "dust" particle recorded in the photographic development when the Lunar Orbiter processed the picture and transmitted it to earth.

4
GLASS SPHERE ENCLOSING THE MOON?

Believe it or not, there is more than one layer of debris artifacts in the process before final printout on earth. There is camera lens dust, micro-fine cosmic dust particles inside the Lunar Orbiter, the tiny straightations or lines in the machines processing of the prints and then there is possibly the scan lines in transmission, not to mention the dust on earth that could have been imprinted on the original copies made for the different Space agencies. You have the NSSDC, JPL and a few others.

After studying the lunar photographs this writer found many other hair-like fine slivers, scratches, blobs, cuts, scan lines and anomalous looking imprints on thousands of frames in thousands of areas on the Moon. Just review one of the many NASA published lunar atlases.

It seems that if Hoagland is right (and he is not), the entire Moon should, for all intents and purposes, have been covered with one gigantic glass sphere. Hmm? Interesting. A glass sphere housing a small planetary "moon" inside! But, it will burden anyone's sanity to find even one humongous colossal support beam column that would have held up such a structure. I can understand maybe the almost complete destruction and decay of the glass dome(s) and even the smaller support beams, but not one larger one has been found! We do see one qualifying as the only possible support beam - the famous Shard, which we shall discuss later. But, it is not big enough.

Needless to say, getting back to Earth and common sense, frame LO-III-85M, which we have been discussing, has uncountable so-called alien hair-like glass crystals, dust particles and processing scratches. We really need not look any further than this frame. This one is enough to keep the bewitched busy for months. There is no need to waste thousands of dollars buying NASA negatives to verify what is being said here.

Getting back to the above optical illusion, the particles or debris artifacts always appear flat to the real surface of the photograph. It seems that no matter which little debris spot one chooses to look at, not one has proper perspective to it. The particles always appear flat to the surface of the photo.

The crystals seem to be found in almost every place you look and much higher than normal for any crystal dome. They can be found even in the corners of the photos, some few off the photo area and others right in the middle. They all seem to run parallel to each other all along the surface. Close inspection leads to other fantastic evidences.

6
UKERT CRATER TRIANGLE

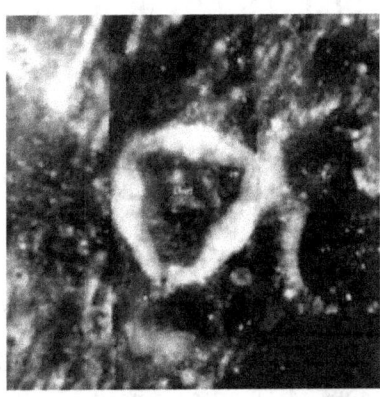

Attached to this array of geometrically shaped objects or what looks like a broken conk shell is Ukert Crater. Ukert is a lunar impact crater that lies on a strip of rugged ground between Mare Vaporum to the north and Sinus Medii in the south. It is located to the north-northwest of the crater Triesnecker and northeast of the crater pair of Pallas and Murchison. The outer rim of this crater is not quite circular, with outward bulges to the north and the east. The interior floor is irregular in places, with a central ridge running from crater midpoint down to the southern wall. It is the one previous discussed as having the little group of UFO scout ships – the cluster of little white blobs. It is a nice little crater not uncommon, but somewhat unlike many others. It has a "tetrahedral triangle" shape inside its crater walls.

Unlike other craters that have various objects, alphabetical letters and multiple rims, Ukert is almost geometrically perfect. It is perfect enough for Mr Hoagland to Heehaw all kinds of occult significances. He scribbles page after page of mysterious mathematical formulas, from Euclidean triangles, to a six-pointed star, the solvability of the quintic, and the trisection of angles, divisibility of instances and theories of alien technology.

Now, after all this pseudo-scientific gibberish we are presented with some simple minded alien baseball diamond shaped field and a funny triangle inside a crater. Obviously, since aliens do not need entertainment nor know anything about sports, we can sterilize all lunar finds of any super hyper-dimensional, cultural entertainment and strictly "sport" the idea of a completely sterile scientific hyper-dimensional purpose. The question now is what does such an anomalous looking square field mean to this alien presence?

CRATER FORMATION AND WEIRD FEATURES

The Ukert Crater triangle and the Diamond Field, along with the busted conk shell fragments are not the only location of odd shaped formations. Other crater gaming fields and odd shaped craters can be found such as the triangular shaped one to the bottom right of Ukert. These natural selenological structures can be found almost anywhere on the Moon. There is a natural explanation for every one of these if the student takes time to think reasonable rather than invent artifices.

The natural explanation of Ukert is crater wall deterioration and collapse, where the higher reflectivity of the lunar soil contrasts with the shadows forming what looks like some artificial alien geometric shape. One of the main images of Ukert is from the Clementine spacecraft. It shows that the sides and the vertices are not linear and as sharp as the Lunar Orbiter photos as Hoagland would have us believe.

This higher resolution photo reveals the triangular anomaly to be natural to lunar geophysical processes. There is nothing bazaar about Ukert except the light-shadow anomaly caused by the combination of crater development shape, high and low reflective lunar soils and sun angles.

A meteorite strikes the lunar surface creating a crater, the walls collapse creating interior crater wall landslides exposing highly reflective materials. In this case the illusion of a triangle is formed by the uneven land slide of only three sides. With the lunar soil of a higher reflectivity, the walls would highlight brighter at lunar Noon-time obviously explaining the supposed artificial looking shape.

The triangle makes its appearance in the optics of telescopes at full Moon (high noon at its location on the lunar surface [9] creating a 16-mile-wide dark equilateral triangle within the crater floor. Spaced 120 degrees apart around the edge are three brighter areas that when connected create the triangle.

Considering the shape and construction of the crater rim and the reflective properties of the crater walls, as well as the brightness of the sun as it rises, it is not difficult to see how the mysterious triangle develops at high noon. The more the light brightens as the sun rises the brighter the three crater rim peaks expand.

Without much consideration of alternatives and a total disregard for real scientific analysis of the crater, Richard fabricates alien "redundant geometry." Obviously a hallmark of his proposed "tetrahedral geometry" -- a two dimensional representation of a three-dimensional, double tetrahedron inscribed within a sphere identical to the geometry found in the Cydonia Complex on Mars.

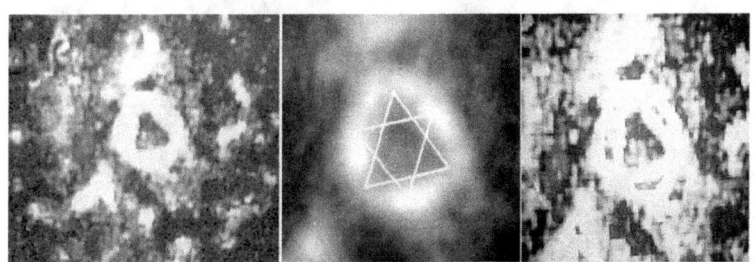

**DEVELOPMENT OF HOAGLAND'S TETRAHEDRAL TRIANGLE
AND SIX POINTED STAR**

6
CRYSTAL DOME OF SINUS MEDII

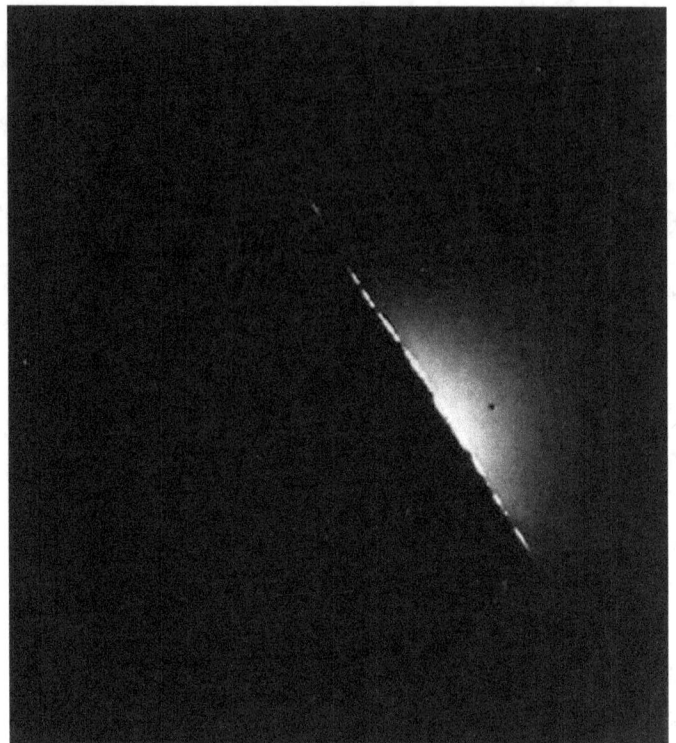

SURVEYOR-6, Photo-7 NASA 67-H-1642
November 24, 1967
Sunlight diffracted at Moon's limb as seen in Surveyor VI picture
of the horizon west of spacecraft.

Richard's search for enlarged pixel distorted crystalline-like structures migrates to the landing site of the Surveyor Missions, particularly Surveyor-6 and subsequently Surveyor-7. The Surveyor-6 spacecraft was the fourth of the Surveyor series to successfully achieve a soft landing on the moon, obtain post landing television pictures, determine the abundance of the chemical elements in the lunar soil, obtain touchdown dynamics data, obtain thermal and radar reflectivity data, and conduct a Vernier engine erosion experiment. It carried a television camera, a small bar magnet attached to one footpad, and an alpha-scattering instrument as well as the necessary engineering equipment. It landed on November 10, 1967, in Sinus Medii, 0.49 deg in latitude and 1.40 deg West longitude - the center of the moon's visible hemisphere.

This spacecraft accomplished all planned objectives and also performed a successful 'hop' rising approximately 4 m and moving laterally about 2.5 m to a new location on the lunar surface. The successful completion of this mission satisfied the Surveyor program's obligation to the Apollo project. On November 24, 1967, the spacecraft was shut down for the 2 week lunar night. Contact was made on December 14, 1967, but no useful data were obtained. [10]

Yet, Richard said, it did send back some fantastic photos of what he believes is an ancient alien crystal dome situated in Sinus Medii. He said of the following photo: *"This may be a photograph of the extraordinary glass dome covering the region of the moon known as Sinus Medii. It was taken by the unmanned Surveyor 6 on November 24, 1967, one hour after sunset."* [10]

This photograph (NSA7-100) of the sun's corona can be seen in the NASA technical report 32-1262, August 15, 1968. The view angle is to the west toward the western part of Sinus Medii. It shows the sun's corona reflecting upwards from below the horizon and through what alien Oopart hunters say is a grid-like glass structure of numerous objects and dome support beams dangling above the surface

Most scientists suggest that this effect is caused by micro particles from the surface trapped in electrostatic clouds hovering over the area, and that when the sun shines through it, it gives off a reflective effect exposing a very light thin electrostatic cloud atmosphere, and a very thin one at that. [10] Mr Hoagland said this is what is left of an alien glass dome.

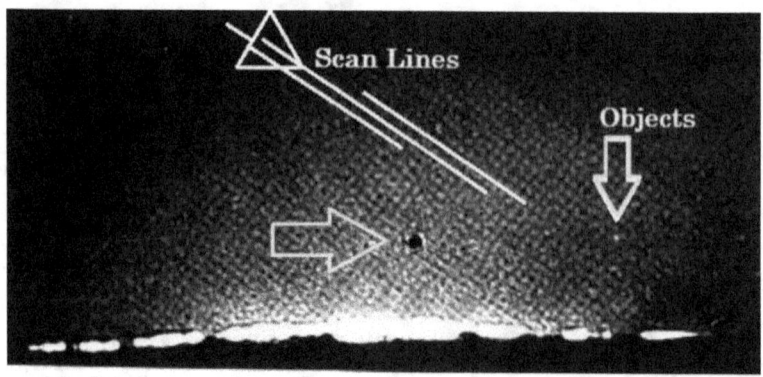

This may be a photograph of the extraordinary glass dome covering the region of the moon known as Sinus Medii. It was taken by the unmanned Surveyor 6 on November 24, 1967, one hour after sunset. - R. Hoagland

Other Surveyor photographs, such as Surveyor 7, seem to suggest the same scientific explanation. Surveyor 7 was the fifth and final spacecraft of the Surveyor series to achieve a lunar soft landing. The objectives for this mission were to: perform a lunar soft landing (in an area well removed from the maria to provide a type of terrain photography and lunar sample significantly different from those of other surveyor missions); obtain post landing TV pictures; determine the relative abundances of chemical elements; manipulate the lunar material; [5] obtain touchdown dynamics data; and, obtain thermal and radar reflectivity data.

This spacecraft was similar in design to the previous Surveyors, but it carried more scientific equipment including a television camera with polarizing filters, a surface sampler, bar magnets on two footpads, two horseshoe magnets on the surface scoop, and auxiliary mirrors. Of the auxiliary mirrors, three were used to observe areas below the spacecraft, one to provide stereoscopic views of the surface sampler area, and seven to show lunar material deposited on the spacecraft.

The spacecraft landed on the lunar surface on January 10, 1968, on the outer rim of the crater Tycho. Operations of the spacecraft began shortly after the soft landing and were terminated on January 26, 1968, 80 hours after sunset. Operations on the second lunar day occurred from February 12 to 21, 1968. The mission objectives were fully satisfied by the spacecraft operations. [11]

MOON MARS MONUMENTS MADNESS

7
SINUS MEDII DOME DEBUNKED

The above Surveyor 6 frame (NAS7-100 photo number 7) of the Sun's corona can be seem in the NASA Technical report 32-1262. [12] The view angle is to the west toward the western part of Sinus Medii. The photo shows the sun's corona reflecting upwards from below the horizon and through what dome hunters believe to be a lunar atmosphere and what they say is a grid-like glass structure of numerous objects (emulsion blobs) and fragmented structures suspended above the lunar surface (See arrows). These anomalous structures and objects are located to the right (west) of the so-called "Tower" and "Shard."

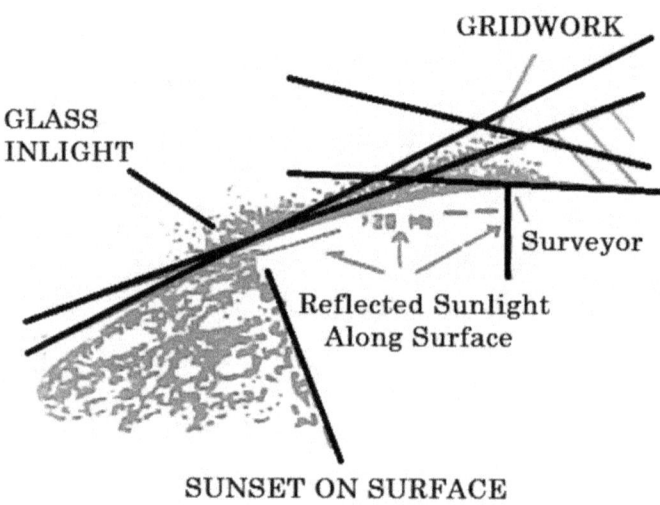

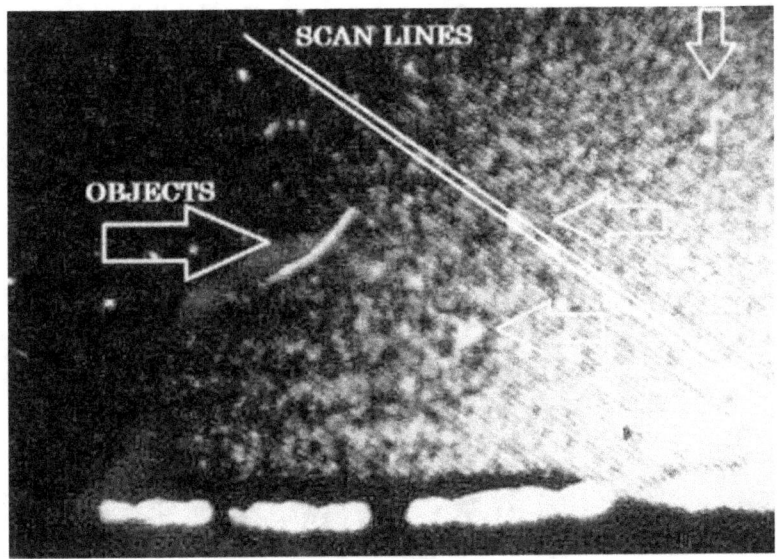

(Left) CHART SHOWING REFLECTION ANGLE
(Right) OBJECTS ON LEFT SIDE

The big difference with this location and series of photos is that they were taken with the light coming from the front from below the visible horizon in what is called *"forward scattered light"* unlike most all other mission photos of the same area, where the sun light is behind the camera causing *"back scattered light."*

In other words, rather than photographing these so-called enigmatic geometric structures in back scattered light, the camera obtained these photos of sun light radiating through some kind of atmosphere in front of and toward the camera. The purpose of these type of shots were to record and study any possible light-scattered properties in the space above the lunar surface caused by the solar corona.

The right side of the photo shows this anomalous reflective material very clearly. Mr Cornet points in his publication that *"This material becomes less illuminated by the sun's corona the further away the structures are located from the axis of the sun and corona, proving its existence."* It is interesting how Mr Cornet replaces the word "material" with the word "structures" leading the reader to believe it is foreign and alien to the Moon.

To the left side of the photo can be seen what appears to be horizontal bands of light reflecting materials and a bright band of bead-like reflective materials along the horizon. Mr Cornet said these anomalies *"cannot be explained as forward refraction from suspended Moon dust."* They, therefore, must be *"some sort of lens-like system of glass-work at the base of a geodesic dome,"* where sunlight is reflecting through rather than off of surface materials. [13]

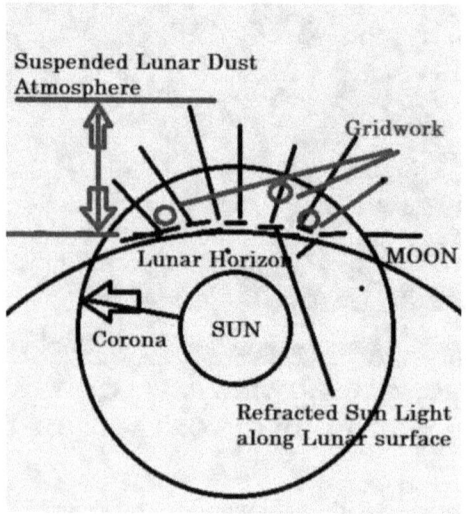

According to conventional NASA sources the view angle of this photo shows intense light from the Sun reflecting off the surface material on the horizon – the bead-like effect reveals the Lunar atmosphere. The grainy-like reflective materials supposedly suspended above the surface and supposedly indicative of some kind of alien built structural dome can be explained by a number of theories.

The large chunk-like objects are probably particles of dust adhering to the camera lens after Surveyor 6 landed. The finer atmospheric refracting crystal-like material is nothing more than lunar dust and other remnant gases of the lunar atmosphere – and the moon does have a very weak atmosphere. The term "atmosphere" is a loose term designating the very minimal gases, particles and electrostatics in the space above the lunar surface. [15]

Nevertheless, Richard Hoagland would have us believe that this material cannot possibly be native lunar dust particles in the atmosphere, because certain other Apollo dust sample experiments eliminated *"suspended lunar dust particles: as a probable cause of this intense forward light scattering effect."* Thus, the Enterprise team of alien artifact hunters concludes that this phenomenon is *"caused by forward-refracted sunlight: myriad lens-like images of actual inner corona of the*

Sun, being literally bent over 20 miles around the sharp curvature of the Lunar horizon... caused by the presence of some kind of glass-like structure."

This of course is only implied and is not proven by independent observations of similar "glass-like" stuff photographed by Lunar Orbiters and Apollo Mission cameras from the same area. There are many ways to explain the floating debris in this photo. Other than the super fine cosmic dust particles floating in the static electricity filled space above the surface there are camera lens dust [15], gases, scan line interferences and the low resolution blurring of the camera. The Sun corona reflects through the lunar dust and gases causing the refractive effect. Then, the image is seen through the camera lens that is obviously less than spotless, the transmission lines and the emulsion patterns when developing the earth based negatives.

Richard asks the question how this reflective effect could take pace in an airless atmosphere implying that the space above the lunar surface is a total empty vacuum and void of all gases. Apparently he denies the scientific reports of the lunar missions that tell otherwise. Gas itself reflects and scatters light.

It was once believed that the moons of other planets including our own moon had no atmospheres. Since the Apollo and other planetary missions, measurements have shown that most of these moons are surrounded by a very thin region of molecules, which can almost be called an atmosphere.

Are there gases on the Moon? Yes, according to recent finds. The weak lunar atmosphere may come from a couple of sources. One source is out gassing where gases are released from deep within the Moon's interior.

The Moon has been found to have abundant amounts of nitrogen, carbon dioxide, carbon monoxide and rare gases such as radon. Another source of atmospheric particles are molecules, which are loosened from the surface when other molecules from space hit the ground. [16] Other noxious gases compose the hot air and cigar smoke within the sitting room of Mr Hoagland and his think tank.

It has been suggested that this Surveyor 6 horizon-glow (HG) is sunlight, which is forward-scattered by dust grains present in a tenuous cloud formed temporarily just above sharp sunlight/shadow boundaries in the terminator zone. Electrically charged grains could be levitated into the cloud by intense electrostatic fields extending across the sunlight/shadow boundaries.

Detailed analysis of the HG absolute luminance, temporal decay and morphology confirm the cloud model. The levitation mechanism must eject more particles per unit time into the cloud than could micro meteorites. Electrostatic transport is probably the dominant local transport mechanism of lunar surface fines. [10]

This would be the lunar version of Rayleigh scattering, which is *"the elastic scattering of light or other electromagnetic radiation by particles much smaller than the wavelength of the light. The particles may be individual atoms or molecules. It can occur when light travels through transparent solids and liquids, but is most prominently seen in gases. Rayleigh scattering results from the electric polarizability of the particles. The oscillating electric field of a light wave acts on the charges within a particle, causing them to move at the same frequency. The particle therefore becomes a small radiating dipole whose radiation we see as scattered light."* [17]

Needless to say, the moon does have enough atmosphere to cause some weak scattering of light, especially when viewed at very low angles over long distances, such as in the Surveyor photos. Here are some more examples or horizon shots showing this effect.

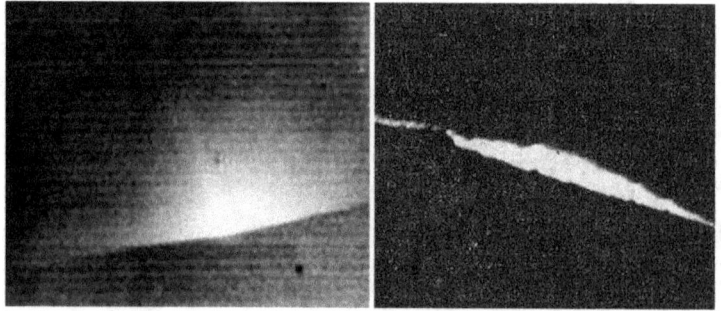

SURVEYOR 1 Solar Corona Spike. SURVEYOR 7 P64b

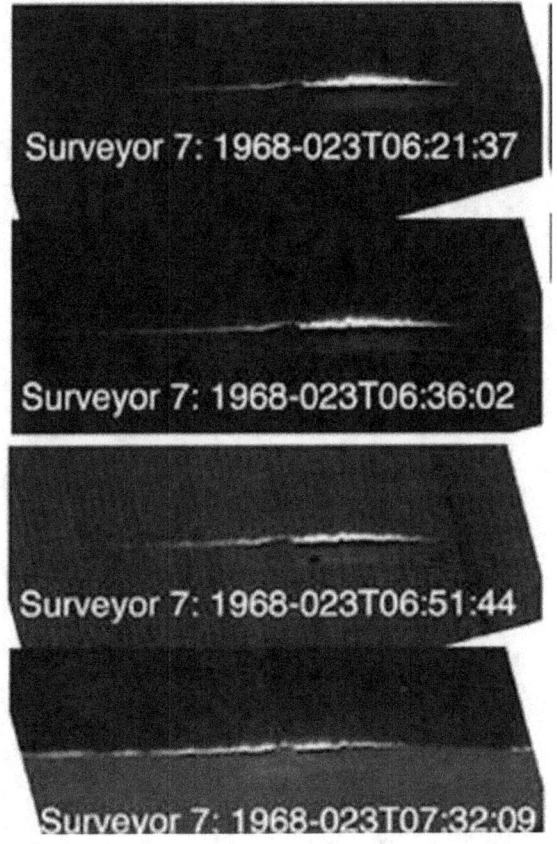

8
THE "SHARD" LOCATION

FAMOUS SHARD PHOTO (Enlargement)
STRUCTURE OF THE "SHARD"
LO-III-84M Famous shot of Ukert Crater area.
LO-IV-101-H3, 102-H1, 108-H3, 109-H1, 12-M

Apollo-10 photographs 4854-4856 show camera looking west at the terminator (Lunar surface sunrise line) from above the eastern side of Sinus Medii. Photo 4822 is looking northeast across Ukert Crater towards the northern edge of Sinus Medii. [See AS10-32-4822, 4854, 4855 and 4856]

The Lunar Orbiter photo II-84M and the above Apollo shots (4854-4856) supposedly show Hoagland's glass-like crystal structures suspended in the Lunar sky along with the alien Tower and Cube mounted on top peaking just above the horizon. The Apollo photos unfortunately for Richard do not show the Tower, Cube nor the Shard and reference can only be made to the LO-III-84M in the following pictures.

Photo 84M was shot by Lunar Orbiter-III towards the south-west side of Sinus Medii from about 30-miles altitude with the horizon at about 256-miles. The Shard and Tower are approximately 230-miles and 200-miles respectively. Orientation of this shot is 45 degrees to the south of the Apollo-10 photos.

LO-III-84M Ukert Crater (left)
Looking towards the south-west side of Sinus Medii.

Other good location photos are LO-IV-108-H3 the Rima Flammarion and Crater Mosting -A area. Locator line "C" passes through four possible locations (A, B, C, and D) for the Shard.

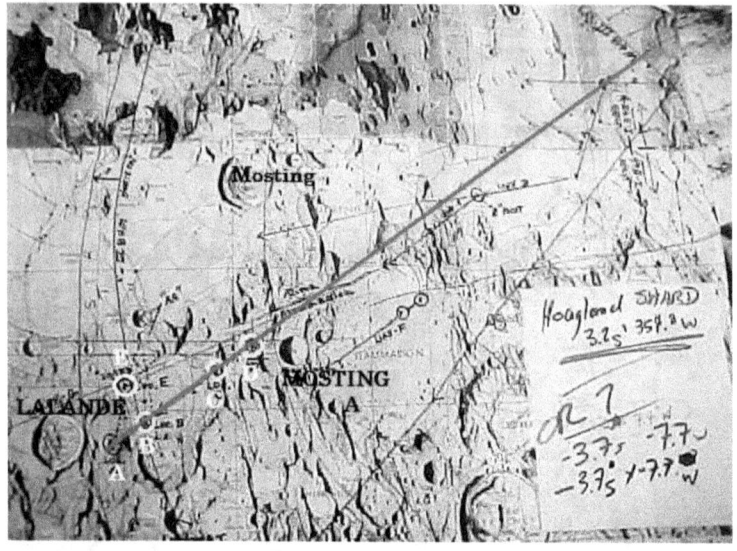

An enlarged and enhanced section of the photo reveals the Shard is an apparently real lunar geological structure sitting out in the middle of the plains area, possibly next to Mosting-A Crater, or beyond westward. As for the Tower and Cube that Richard mentions, it is identified as a faint emulsion blob splatter stemming from the huge splatter line running across and above the horizon. Many of the blobs are compared to show a similarity between the blobs and the Tower Cube. It is most probably an emulsion blob as it too demonstrates the same exaggerated pixel "crystalline structures" as the other emulsion blobs.

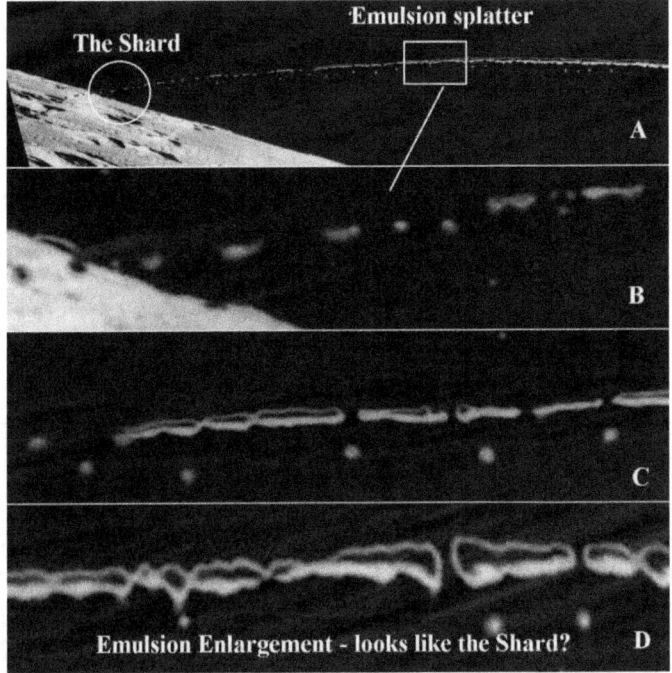

EMULSION SPLATTER BLOBS

After analyzing detailed enhancements of the Tower, Cube and other anomalous objects for so-called "suspended glass-like structures," "support beams" and "alien architectures" all that can be derived from eye torturing study are developmental emulsion splatters, blobs, scratches, smears, and (with Hoagland's photos) over enhanced, digitally exaggerated pixel distortions. In fact, almost every Lunar Orbiter photo of a horizon background can be over exposed, extremely enhanced beyond normal enhancement and made to expose so-called alien dangling crystal structures. LO-III-84M is loaded with them from top to bottom.

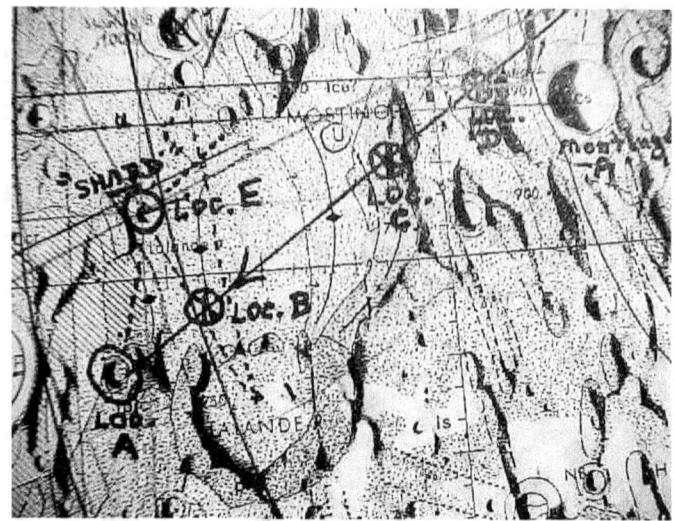

Location Lines A, B, C and D show possible Shard locations.
"A" = 4.5 degrees S, 7.7 degrees W – Hill Top

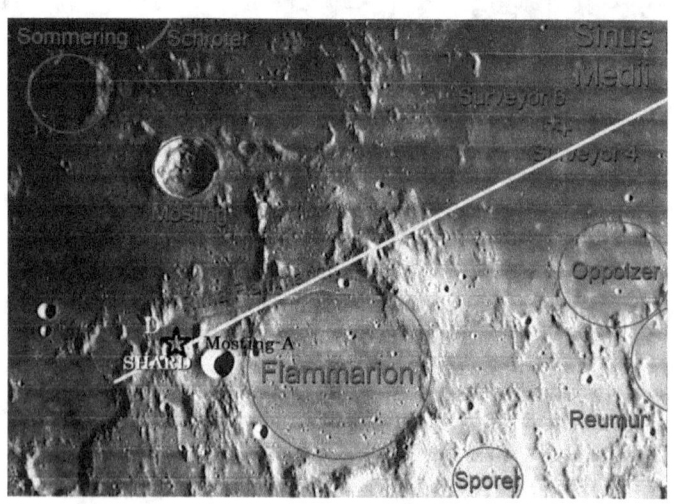

Shard Location "D" is closest to the camera at 3.1 degrees S, 5.8 degrees W, next to Crater Mosting-A

9
THE "SHARD"
(LO-III-84M)

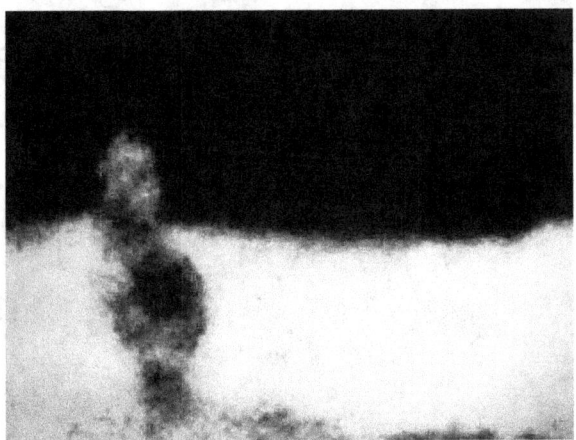

The "SHARD" (Super enlargement)

There are a number of other photos of Sinus Medii besides Surveyor 6 NAS7-100 of special interest relative to alien structures. There are those of the Lunar Orbiter Missions [18] and those from the Apollo Landing Missions [19], especially the famous "Shard" photo LO-III-84M. Sinus Medii is one of the most interesting sites for Mr Hoagland as it is supposedly the location of a great dome and many alien structures. He has found, along with the Shard, such colossal structures as the "Tower" and the "Cube."

LO-III-84M Sinus Medii Horizon

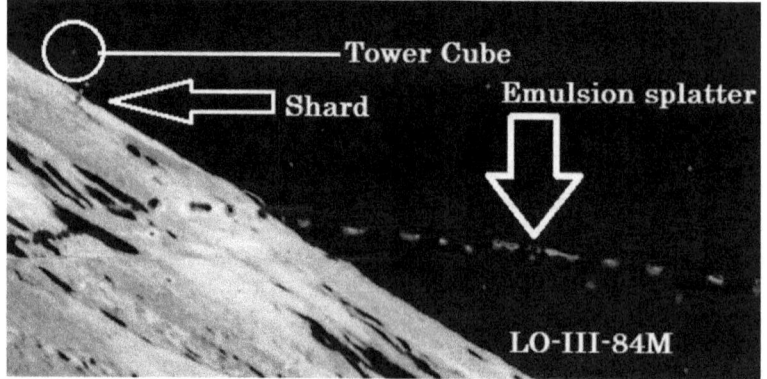

**LO-III-84M Shard, Tower and Tower-Cube
Compared to emulsion splatter and emulsion blobs**

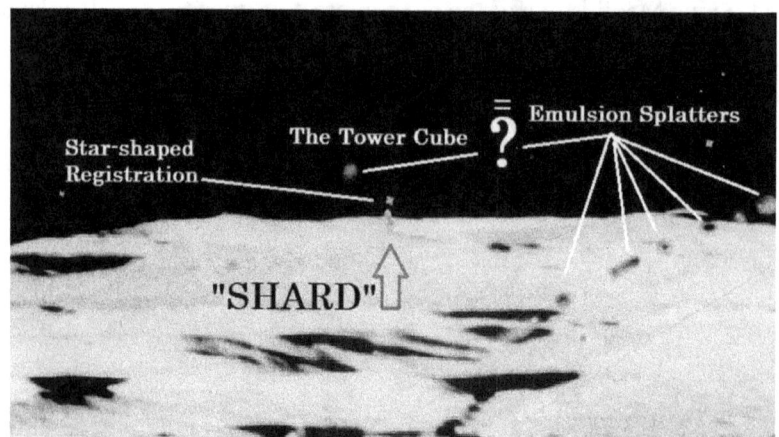

Shard, Tower, Cube and Emulsion Splatter Compared

The Shard is a very obvious tower-like structure that rises above the moon's surface to a possible height of over one and a half miles. It has an irregular *"spindly shape"* of regular *"geometric patterns"* and has constricted modes and a swollen internode.

Mr Cornet as well as Richard Hoagland points out that no known natural processes so far observed can explain such an object of that shape and size. As far as the construction of the Shard, its high reflectivity exposes its artificial nature and alien manufacture. *"The amount of sunlight reflecting from parts of the shard indicate a composition inconsistent with that of most natural substances. Only crystal facets and glass can reflect that much light..."* - i.e. on the Moon. The shard must therefore be *"a highly eroded remnant of some sort of artificial structure made of glass-like material."*

The Shard is an amazingly huge object but not really made of some amazingly artificial alien material. The reflectivity of the object when compared with surrounding reflective lunar materials reveals that it is not much different in composition than the Lunar regolith "soil" and no more "unnatural" than any other object in its surroundings.

It is not all that odd in shape and size, though it is unique in its position and placement on the surface. Seems it is the only large object in the area. Even Mr Cornet points out other large structures in the area with the same reflectivity, though not as large as the Shard.

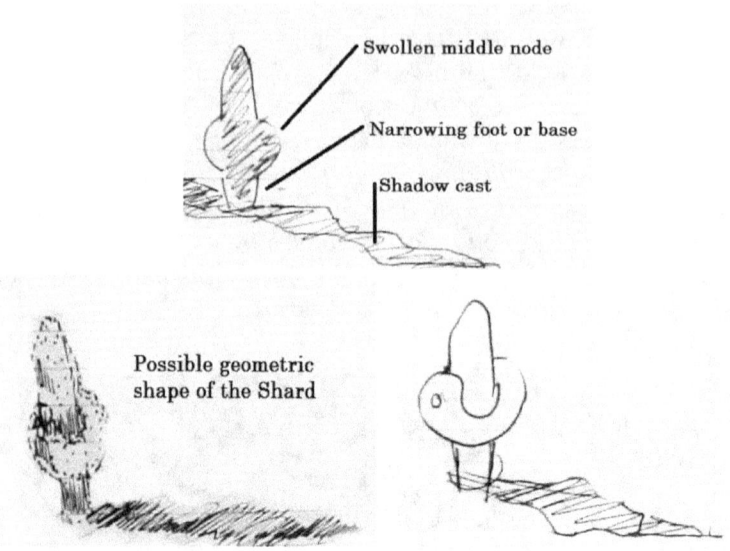

The Shard is shown casting a shadow revealing that this object is without a doubt real. The shadow mirrors to a great degree the Shard's unique shape. The shadow is also consistent with the local surface geology on which the shape is sitting. Hoagland said, "*It is this shadow – more than any other aspect of this object – which in the eyes of many observers solidifies its reality as an extremely anomalous, manufactured lunar structure.*" [Martian Horizons V2, #5. p.16] This is one of the few things Richard and Mr Cornet are right about. The odd size and shape could be explained away as an emulsion blob if not for the shadow. The shadow forces one to admit it is a real lunar surface object.

According to these two guys, the size and shape, in association with the shadow "*argues forcefully for an origin far more recent than the extremely ancient, rounded and eroded lunar landscape that surrounds it.*" The shard, they continue "*somehow has managed to defy the laws of entropy that have determined those surroundings (according to geochemical and isotope age data returned by the Apollo missions): an incessant meteoric rain, which, in the absence of "recent" geologic activity, should have reduced this object to rubble indistinguishable from those of its surroundings.*" [p. 17-18]

Mr Hoagland, then, continuing on page 18, rattles off super

geometric jargon describing the advanced crystalline structure of the object. Apparently, he over developed and over ultra-enhanced the shard to the point the pixels degrade into geometrical shapes, lineation's and cubical shapes revealing *"highly-eroded, highly reflective internal crystalline geometry"* with *"details of optically active vertical lineament patterns."* [p.18]

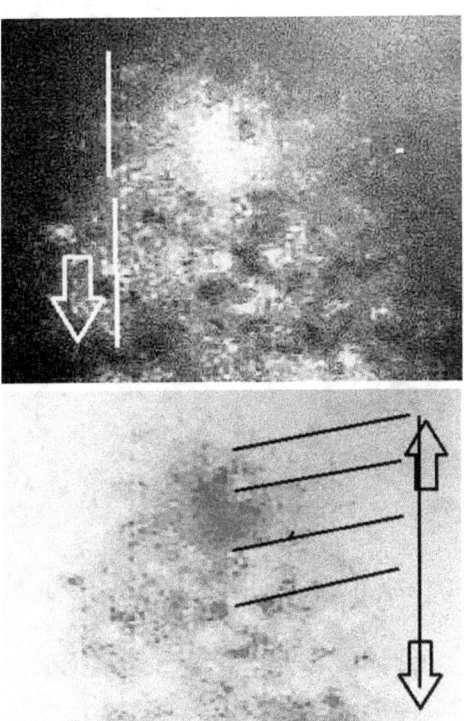

**Extreme enlargement of Shard with vertical lineation pattern.
Could be the pixel pattern?**

10
"TOWER" AND "CUBE" LOCATION

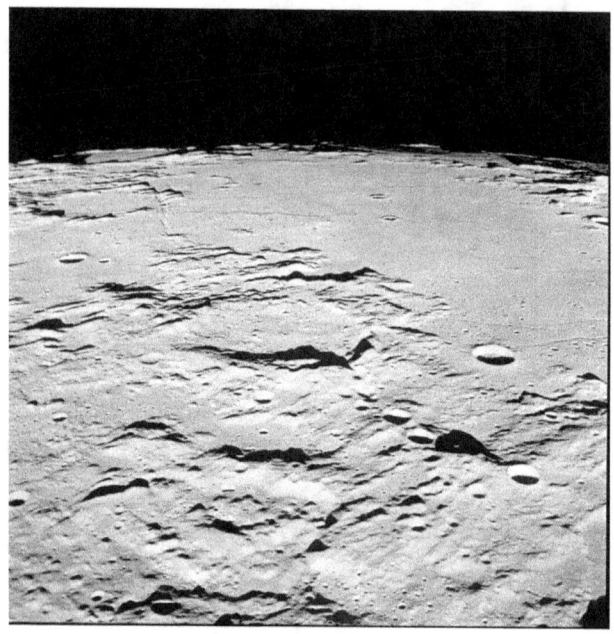

AS10-32-4855
APOLLO-10 photo looking toward Sinus Medii

The tower and Cube are seen to the left of the Shard in 84M. It looks like another emulsion blob, but Richard Hoagland said it *"appears at first to be merely a 'smudge' on the original full-frame NASA photograph – with a 'tail' extending down toward the lunar horizon."* But, it is not any other possible thing – it is not a nebulae, a gas cloud, a comet, dust, nor a distant galaxy, or any other variety of mundane deep space objects- but *"a massive 'mega-cube' hanging more than 7 miles above the Moon!"* [Martian Horizon p. 19-20] Wow! Seven miles above the lunar surface? This is nothing. Wait until we discuss the Castle structures that hang 35-150 miles above the surface close to Ukert.

The above photo AS10-32-4855 and the following 4856 frame show a shot looking westward of Sinus Medii over Rhaetieus, Reaumur, Oppolizer and Flammarion-T at Flammarion and to where the so-called Tower-Cube just barely peeps above the lunar horizon. It is hidden among all the other small tiny specks of defects within the photo. If one looks careful they will see a tiny white dot – the Cube.

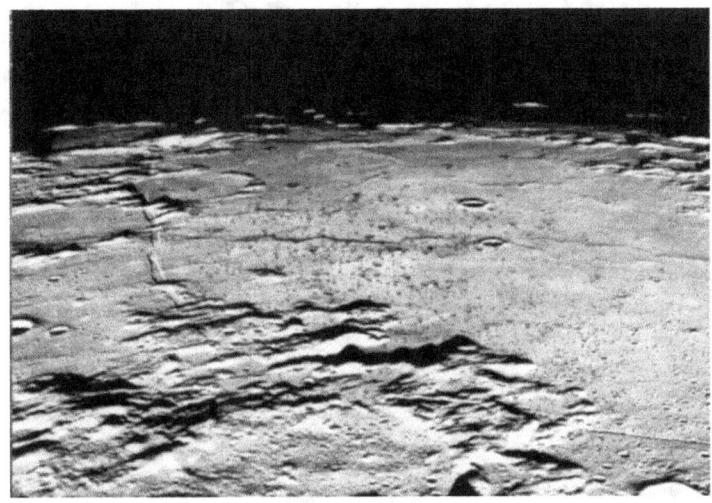

AS10-32-4856

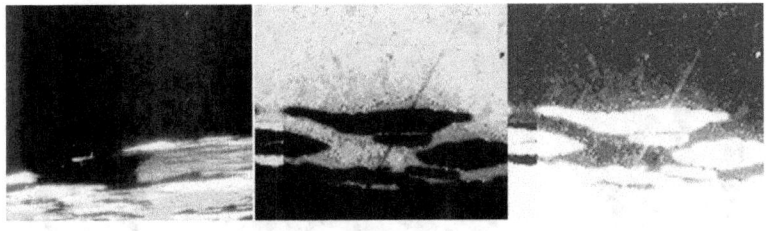

    4856   4855a   4855b

Ultra-enlargements of Sinus Medii area of Shard, TOWER and CUBE locations

AS10-32-4856
Can you see the top of the Tower and the Cube? Take your best pixel pick!

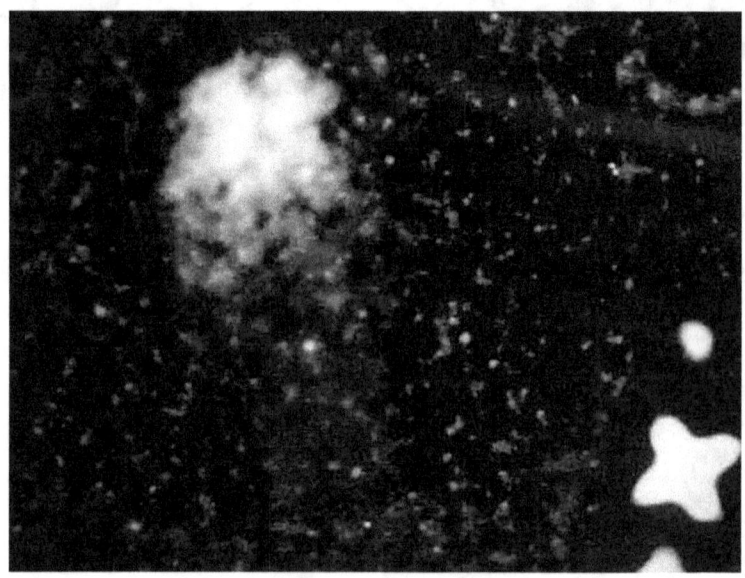

SHARD photo 84M
Tower "Cube" to left of cross-hair star and top of shard

AS10-32-4856
The CUBE

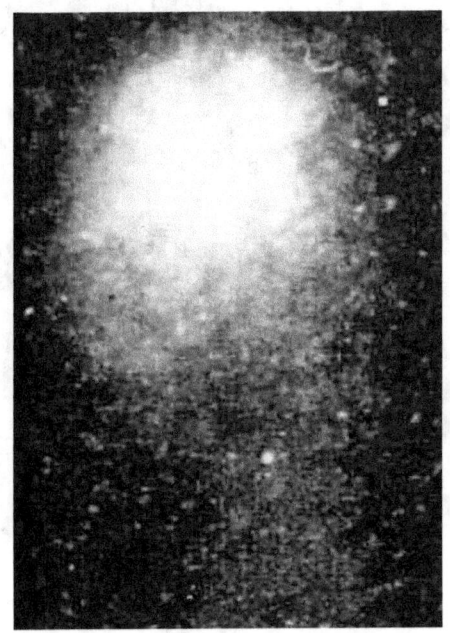

LO-III- 84M [1]

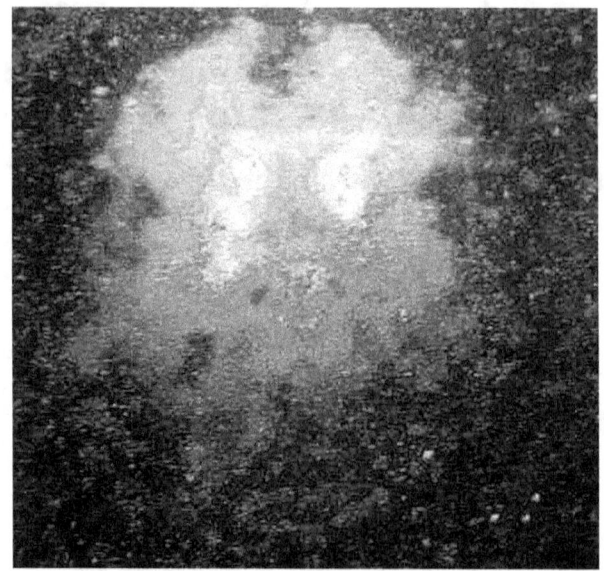
LO-III- 84M (2, 3, and 4) MORE CUBE (enhanced)

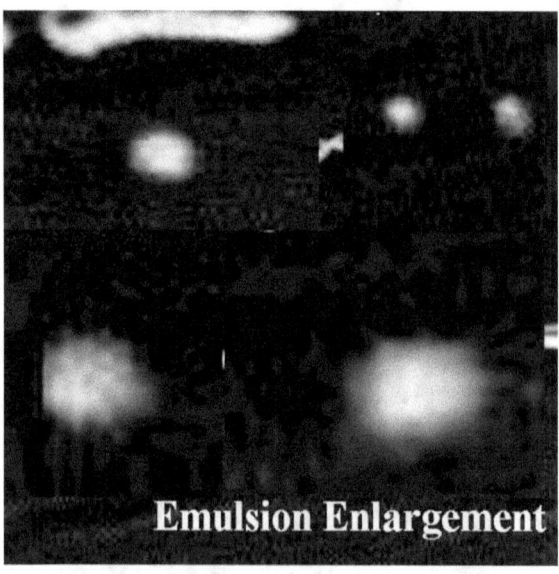
Emulsion Enlargement

11
WHERE IS THE "SHARD" ON USGS LUNAR MAP I-566 ?

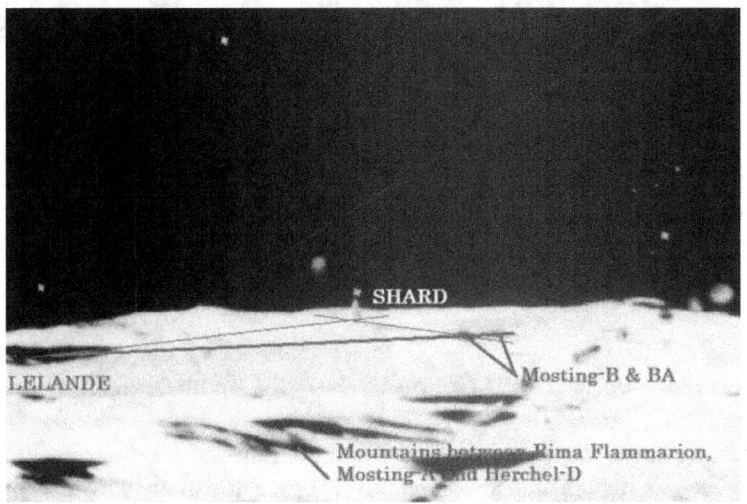

Where in the lunar world is the Lunar Orbiter III-84M "Shard" located? This was one of many questions in my lunar surface studies that I found to be lacking in the Hoagland Moon-Mars Connection conference. And it was one of the most difficult questions to answer. The results revealed 3-4 possible locations.

Looking at the USGS Lunar maps of Sinus Medii area (Maps I-566 and I-548), I found myself at a loss when I first approached the question. Where was I to begin? What I needed were some location "beacons" for locating the Shard. I needed shadows, craters, boulders, mountains and other prominent geographical features for triangulating the location the Shard. The answer was found in the same "Shard" photo LO-III-84M.

To locate the Shard, which was centered in the upper left corner of 84M along the horizon, I needed to identify as many of the craters and surface features as possible. These would be used to sight the Shard's longitude and latitude. The photo "footprint" map in the NASA publication SP-200 "The Moon as Viewed by Lunar Orbiter" helped me identify the photos and the areas covered by 84M. Along with the footprint map, the USGS maps helped to locate and identify the 84M craters and features. The footprint map helped to identify the following 84M area covered: III.84M Parameters: Blagg 1.1N. X 1.3E - 2.5N. X 0 degrees. Flammarion 2.4S. X 1.4W. (Random point) - 2.5N. X 7W. (Random point)

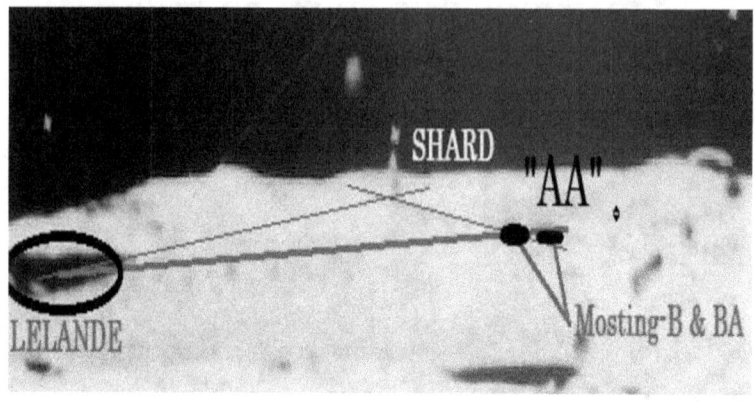

The geometry runs as follows: The 84M photo horizon appears to extend to Crater Lalande in the area South-west of Sinus Medii. Checking the 84M photo reaffirmed the extent of the horizon and the parameter in which the Shard rests. To begin locating the Shard, while drawing lines on the 84M photo and the USGS Maps, the next step was to locate and identify Crater Bruce as a beginning point to run lines out toward the horizon to pin-point the Shard on the USGS Map. This was easy. I drew a line (D) from Bruce to Crater Blagg which extended the first sight line to the horizon past Crater Mosting.

Another line (A) was drawn from Bruce to Crater Oppolzer-A (0.3S. X 0.2W.). From (A) was drawn a line outward towards and through a little mound (Z) north and north east of the Flammarion area (0.5 degrees S. X 2.0 degrees W.). My "E" line was extended further to a small crater (Flammarion-C) on 84M and the Map in the North of Flammarion (2.0 degrees S. X 3.5 degrees W.). The crater's location actually begins what is named the Rima Flammarion, a fault line which runs towards the Shard!

One can see on 84M the Rima Flammarion passing to the right of the Shard, where the Shard rests to the left on a small plains area or on top of a hill--Sighting the Shard somewhere between Lalande-R, Lalande Crater and the two small crater lets (which I name "AA", as circled on the photo blow-up of 84M (Mosting-A & BA., 2.4S. X 7.1W.). I thus sighted the location of the Shard somewhat beyond "Location-D" (3.1 degrees S. X 5.4 1/2 degrees W.), possibly at map Location-C (3.3 degrees S. X 6.0 degrees W.), or somewhere along my line from Location-C to B (4.1 degrees S. X 7.1 degrees W).

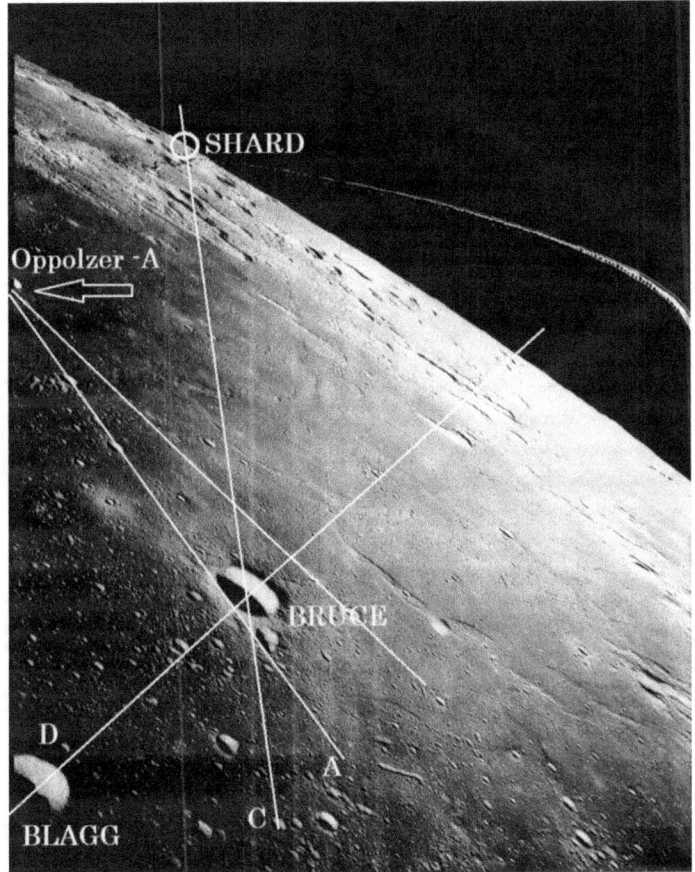

Triangulation to locate Shard

The most probable location for the Shard was found to be one of two prominent little mounds sitting to the east of Crater Lalande at Locations-A and E (4.3 degrees S. X 7.4 degrees W., and 3.5 degrees S. X 7.3 degrees W.). The Location Line-C, running from Bruce S.E. through Flammarion and past Mosting-A Crater, therefore runs through Rima Flammarion to Crater Lalande. The Shard, according to the 84M crater location sight lines, is thus located somewhere between Mosting-A and Lalande in a small area North of Lalande-R, somewhat north from Line-C. If one observes the little 84M photo (blow-up) at the end of this section, it will be seen that the Shard sits away from and beyond the double Crater "AA" or Mosting-B and BA, and Crater Lalande-C.

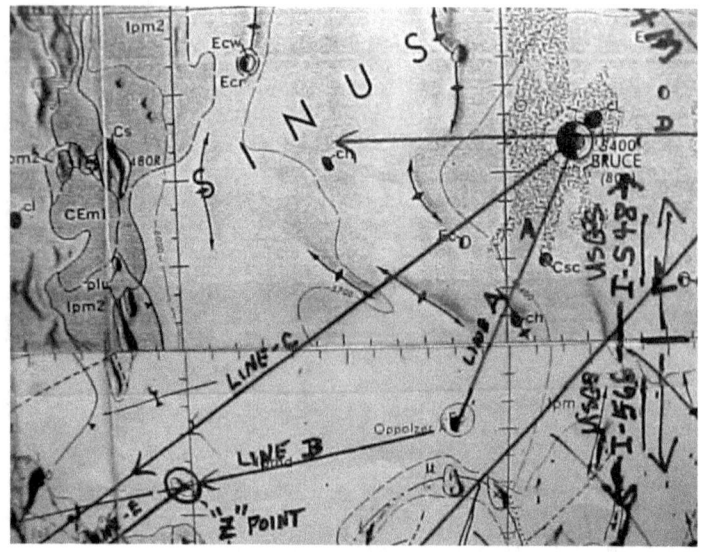

MAP-A

Behind it, according to the Map, sits a small scarp or hill. Past this is the big Crater Lalande. To the left there is a small horizon hill which is Location-A. The Shard thus sits at Location-E at 3.5S. X 7.3W. From Bruce to the Shard along Line-C, according to the USGS Map, is about 170 miles.

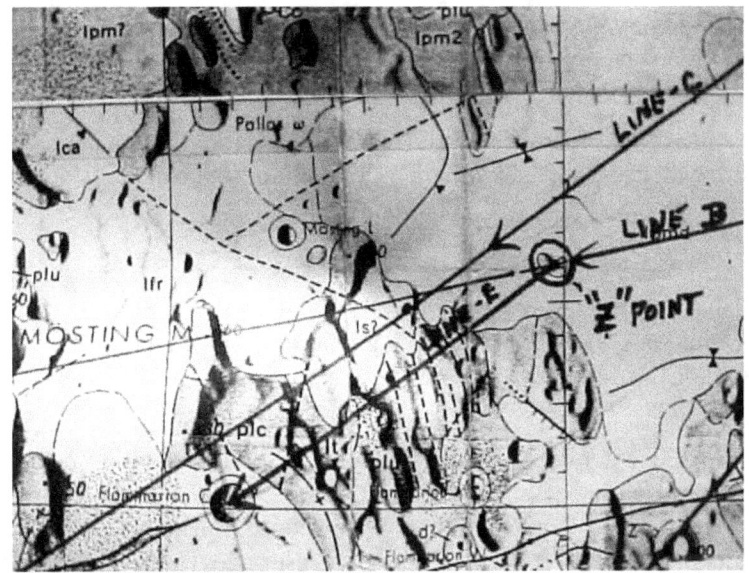

MAP-B

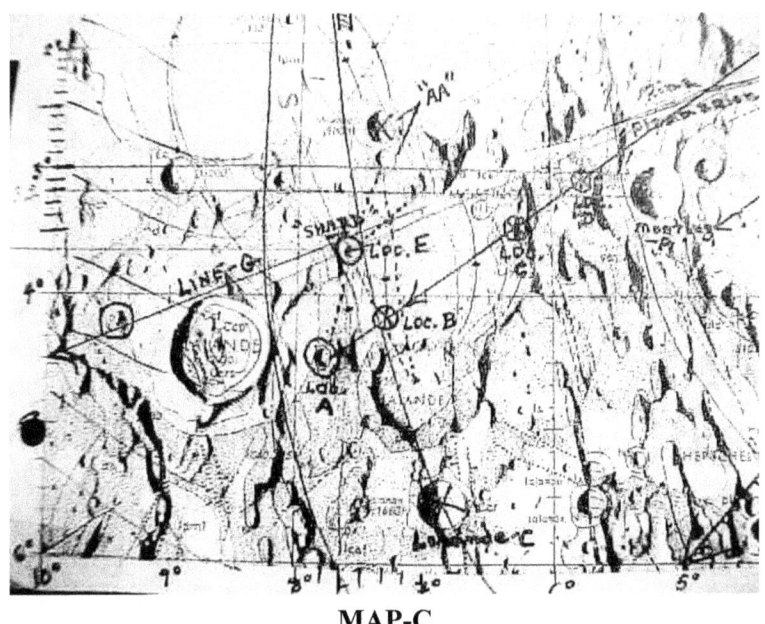

MAP-C

MAPS AND PHOTOS:

A. USGS Maps (I566 and I548) of Sinus Medii area with positions of locations and Shard.

B. Negative Print of LO-III-84M "Shard"

C. Negative Blow-up of LO-III-84M with location lines and identified craters with Shard in background. [Dotted line in sky is emulsion error.]

Crater Lalande-C F

Location-A F

E. Mosting-A & BA. My "AA"

G. Mountains between Rima Flammarion and Mosting-A & BA., Mosting-U on Map.

The Cube-Tower, which lies beyond the Shard, is still to be located. The problem in locating the Tower, according to 84M is that it rests beyond the 84M horizon, somewhere along Line-G. [Check out the mound at 4.1S. X 9.3W.] The lack of surface features for sighting presents a problem. Perhaps there will be other photos taken that will help locate the Tower?

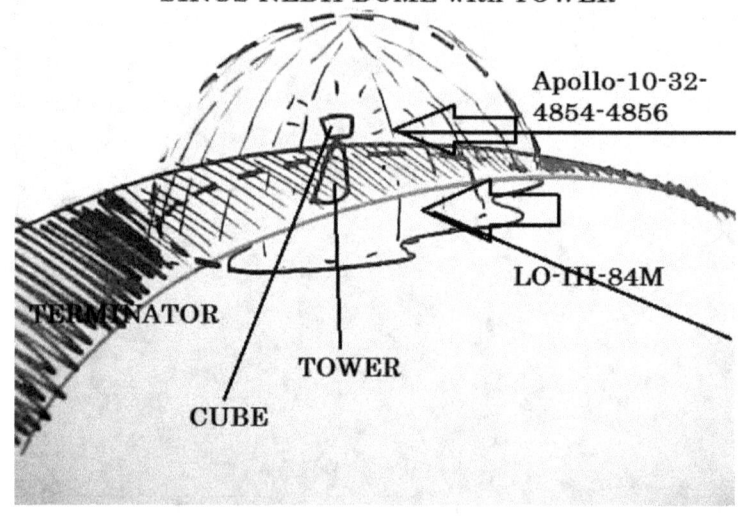

SINUS NEDII DOME with TOWER

The Missions numbers that I have compiled that cover the Lalande "Shard" area are, L.O. IV 113H, 114H, 108H and 109H. The maps covering the Sinus Medii area and the location of the Shard are, USGS Lunar Maps: Macrobius Quadrant I-799 (NW), Cleomedes I-707 (NE), Taruntius I-722 (SW) and Mare Undarum I-837 (SE).

As mentioned before, there are four possible locations for the Shard: Location A, B, C and E, discounting D. Locations A and B are located respectively 4.5S x 7.8W and 4,2S x 7.4W. Location C is at 3.5S x 6.3W. "A" is situated directly in-between Craters Lalande and Lalande-R. Shard position B is along line "C" at about 11 o'clock to Lalande-R. Location C is north-west of these at 3.5S x 6.3W, west of Mosting-A sitting on a large hill close to Mosting-U. This is south-west of Richard's specific location 3.2S x 354.8W, which I cannot find.

Possible location E is a very interestingly accurate site for the Shard. If one takes LO-IV-113-H3 and compares it with 84M (see chart) a very small crater will be found between Lalande and Mosting B/BA. Location line G runs through this point. Notice that the shard is a 1 and ½ mile high artifice and not a crater. In reviewing all the data and photographs, not one revealed any shard, tower or cube.

No alien shard, skyscraper or cubical shaped tower hotel ever jumped out from the frames except in 84M. Nothing. No clues, no peaks or parts or suggestions ever appeared. 84M frame is a lone ranger, the one and only photograph with the Shard. With the shadow it is casting, it appears so real and natural to the lunar surface, yet only in this one picture. Therefore, it is suspect to artificiality and photo processing development error and not being natural to the moon. It is not conclusive that it is real since it is not in any other photograph and the one photograph that it does appear in is choked with emulsion blobs and splatters.

According to the enlargement of the shard photo 84M, the shard appears situated behind and approximately in-between Lalande-R and Mosting-B east of the invisible Lalande Crater at about 3.7S x 7.7W at point D. Obviously, the first estimated locations A, B and C lineation's and cross references outlined on the map failed to truly locate the shard. Richard's location for the shard at 3.0S x 354.8W also fails as well to fit the surface details around the shard in 84M. His location places the shard too far into the foreground.

Accepting Hoagland's location "D" would place the shard Mosting-A, Flammarion-D and Mosting-B on the map. Yet, location E befits the LO-III-84M surface features and places the shard at 3.7S x 7.7W.

As mentioned before, none of the other possible photos of the area revealed any Shard. Looking for locations A and B in such frames as LO-IV-113-H3 is fruitless. Comparing craters Lalande and Lalande-R with the 84M revealed no shard. Also, none of the Apollo photos showed any super high structure in this area either. [20] Location C was blank as well. LO-IV-108-H3, 109-H1 photos of the Mosting-A crater area also revealed nothing of the shard as found in 84M.

What of the Flammarion area around Mosting-A? Would some of the other photos taken from different views reveal the shard? Search until your eyes cross. There is no shard to be found. I need not display any photographic proofs when there is no evidence to be found. Anyone can check these locations on the Internet for themselves.

So, what is the conclusion regarding the existence of the shard? Is it truly there as 84M seems to show? This one and only photo does show an object and it seems to cast a shadow suggesting that it is really present on the Moon. As for any other evidence, no other photos seems to support the object.

There are many reasons why this is the case. The angle is wrong, the details are vague and the lighting is either too dark or too bright, and the object is too far away to be seen. 84M seems to be the only photo of the shard. Maybe the LROC (Lunar Reconnaissance Orbiter camera) will reveal something? [21] So far the LROC has not mapped this sight yet.

12
LUNAR RECONNAISSANCE ORBITER CAMERA PHOTOS

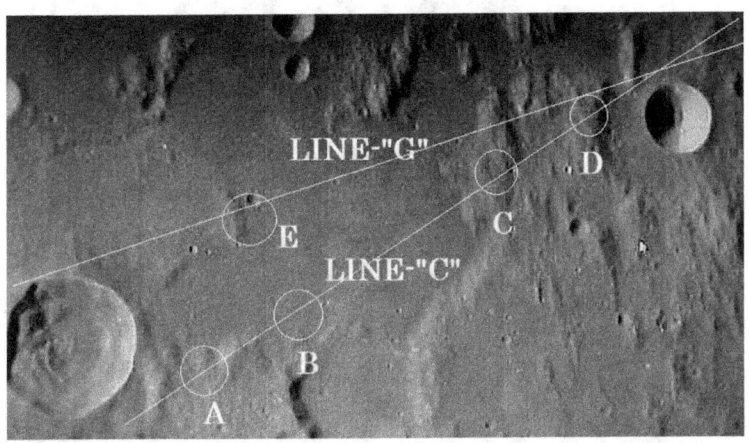

LUNAR RECONNAISSANCE ORBITER "Shard" Locations
[http://target.lroc.asu.edu/q3/http://target.lroc.asu.edu/q3/#]

The shard should be somewhere between Craters Mosting-A and Lalande, and around Lalande-R. Remember, this wild goose is supposed to be a 1 and ½ mile high tower like shard. Crater Mosting-A is about 11-12 kilometers or approximately 10 miles wide. Therefore, the 1.5 mile high shard should stick out like a sore eye somewhere between Craters Mosting-A and Lalande and above Lalande-R in the above enlarged area photograph. The Lunar Reconnaissance Orbiter photographs seem to be void of any alien manufactured super structure.

Location - A Shard Missing

Location - B Shard Missing.

Location - C Shard Missing

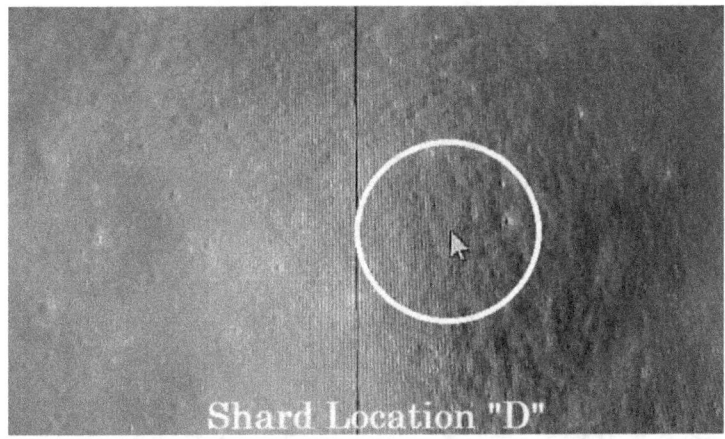

Location - D Shard Missing

Location - E Shard Missing

There is no tower, no shard, no playground park or swings, no swimming pools, no nothing! Nevertheless, maybe there is some other unallocated spot for the shard? - Have fun looking! And when you get tired, return to LO-III-84M for the only relief available. Could the shard also be just an oddly placed emulsion blob or splatter mark? Your guess is as good as mine or Richard Hoagland's. As far as this writers is concerned this mysterious alien artifact is just another Richard Hoagland charade.

POSSIBLE LOCATIONS:

Bruce..1.1N X 0.2 1/2E
Oppolzer-A...0.3S X 0.2W
Z-Point..0.5S. X 2W
Flammarion-C (Line-E).............................2S X 3.5W
Location-A (Lalande Mound)....................4.3S. X 7.4W
Location-B (Possible Shard Location).....4.1S. X 7.1W
Location-C (Hill Top)................................3.3S. X 6W
Location-D (Hill Top)................................3.1S. X 5.4 1/2W.
Location-E (The Shard?)...........................3.5S. X 7.3W.
Mosting-A & BA...2.4S. X 7.1W.
Shard's Most Likely Location. A little mound east of small scarp-hill ....3.5S. X 7.3W.

13
REBAR BEAMS OF CRATER MANILIUS
AS10-32-4822

Shot (enlargement) of Crater Manilius to the far right of Ukert
LO-III-84M

In looking for more evidence of crystal domes and alien structures and habitations, Mr Hoagland super enlarges a crater area in the top right corner of Apollo-10 photo 32-4822. The little crater is Manilius and is located about 7.5 N x 1.5 E on the USGS Maps. Our two-dimensional photographic forensic fanatic points out what he believes are *"extraordinary structures"* such as bridges, suspended crystals and rebar beams hanging around the atmosphere above the crater.

In observing closely around and above the crater, one will notice what looks like black streaks or thin multi-stranded features with some having three repeating parallel bands blatantly caught looping across the upper right hand top corner below the crater.

These extraordinary *"extraterrestrial suspension bridges"* arch several miles above the lunar surface. According to Mr Hoagland this is a *"prominent, 3-D sagging, three-stranded 'lunar-bridge' appearing in silhouette."* He continues to comment that on close inspection the photo shows what appears to be a series of right-angle *"brackets,"* parallel *"couplings"* (A) and a *"box-like framework"* suspended from the span (B). [22]

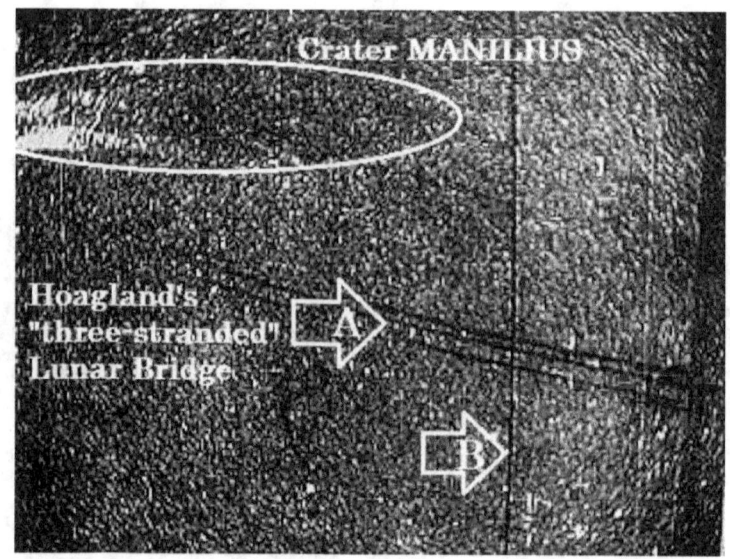

Crater Manilius and the Sagging Bridge

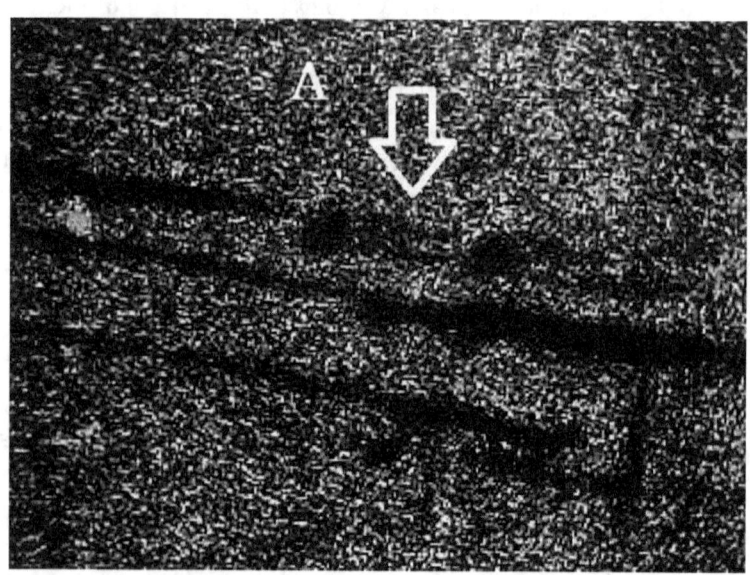

Close-up of three-stranded Bridge (A)

Our bridge builder goes on to propound that these *"artificial-looking structures, these remarkable 3-D details are increasingly difficult to* explain as any type of 'natural lunar feature'... [Or as] any 'trivial explanations – such as 'scratches on the original NASA negatives or prints." So, he asks, what are these "*mile-long, stranded, lunar structures,*" "*sagging cable cars*" or "*hanging lunar bridges*" doing still clinging to equally incredible "*shimmering glass-like lunar towers*" - hanging miles above the Moon? He said, they are "*definitely not recognized, even in NASA, as natural geologic features.*" Of course not, they are emulsion scratches!

On even closer inspection, one wonders where Richard gets such amazingly imaginative ideas. Are these black gouges in the photo really suspension bridge cables or actually scratches doctored over with black ink? There are only three choices to choose from. First, they are photographic scratches. Secondly, they are real suspension bridges. Thirdly, they are natural geologic features.

Which of these is the true explanation for these anomalous features? The third has been universally agreed upon as not being the case. This leave the other two choices. The first is obviously the conventional explanation and is discounted by Richard. The second is discounted by NASA but propounded as the truth by Richard.

Since Richard forgets to point out or purposely overlooks the fact that the larger suspension bridge cable (running counter and vertical to the three-stranded cables) runs the full length from way above Crater Manilius to way down below this area on a perfectly flat vertical plain – like someone took a knife and dragged it from the top to the bottom of the frame across the surface of the photo, we conclude that these cables are really emulsion scratches. Further inspection supports this as there are other "black cables" in other areas of the photo, in such odd places as to suggest that not even aliens would be so stupid as to put them there.

Another scratch appears 2/3's the way down, horizontally parallel to and to the right of Ukert Crater and to the far right in the photo. After some minor studies of these scratches, it is found that over 98% of the photo has these emulsion errors running vertically from top to bottom. The "black bridges" or suspension cables are scratches retouched and corrected with black ink by NASA technicians to salvage the photo for publication purposes. It is too beautiful a photograph to leave hideously scared. In fact, it is the only one of its kind unlike any other photo taken of the area. The filth, dirt, dust and scratches are common to most all NASA photos, with this one being effected more than others.

There are many more of these black lines. When the crater area is enlarged the inside of the crater presents a pattern. It also presents even more microscopic debris, hairs and scratches. Processing lines are visible within the yellow brackets. When the area around the crater is enlarged there can be seen many more black cables running oblique to the horizontal and vertical plains. Others (processing lines?) run vertically from top to bottom. Yet, in all the other Apollo 15 and 17 mission photographs of the crater the area shows no such "extraordinary" structures dangling in the upper atmosphere.

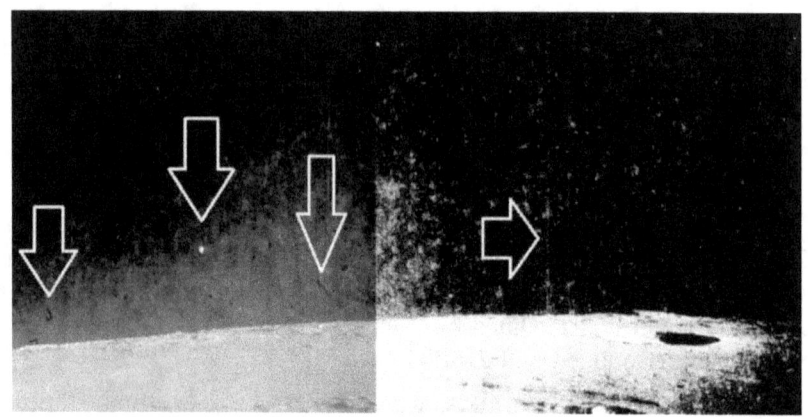

Processing lines, emulsion scratches and debris, hair and dust particles

More debris, hair and dust particles

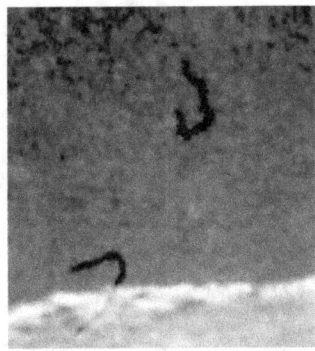

Example of "hair" from dust particles

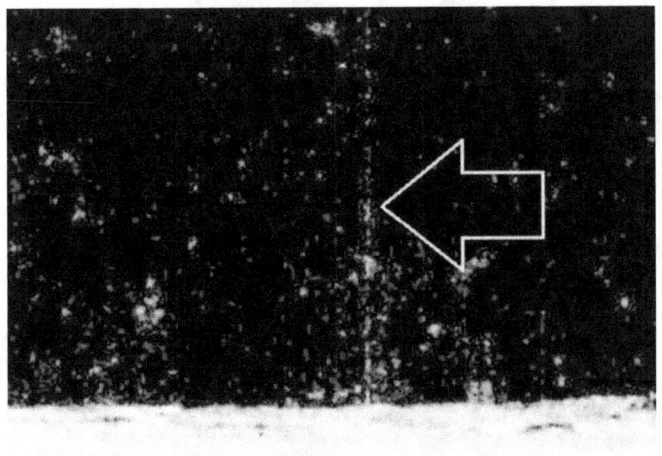

Processing lines

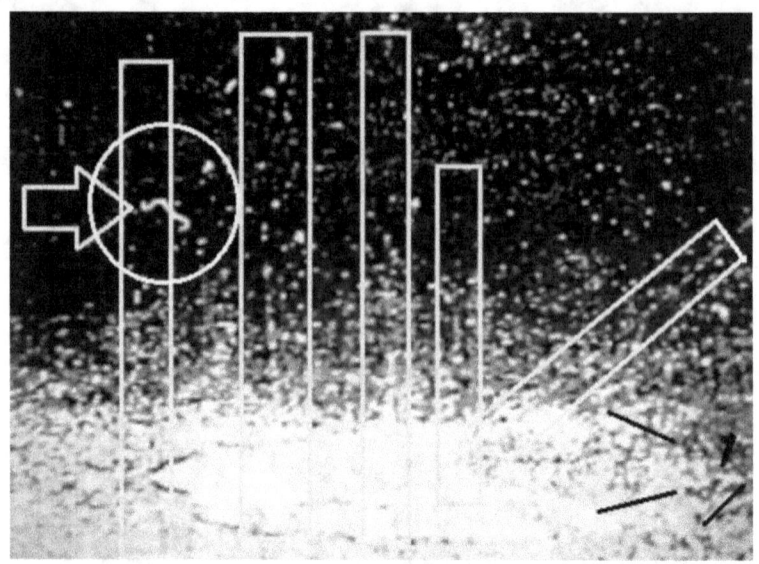

Processing Lines and blotches (enhanced)

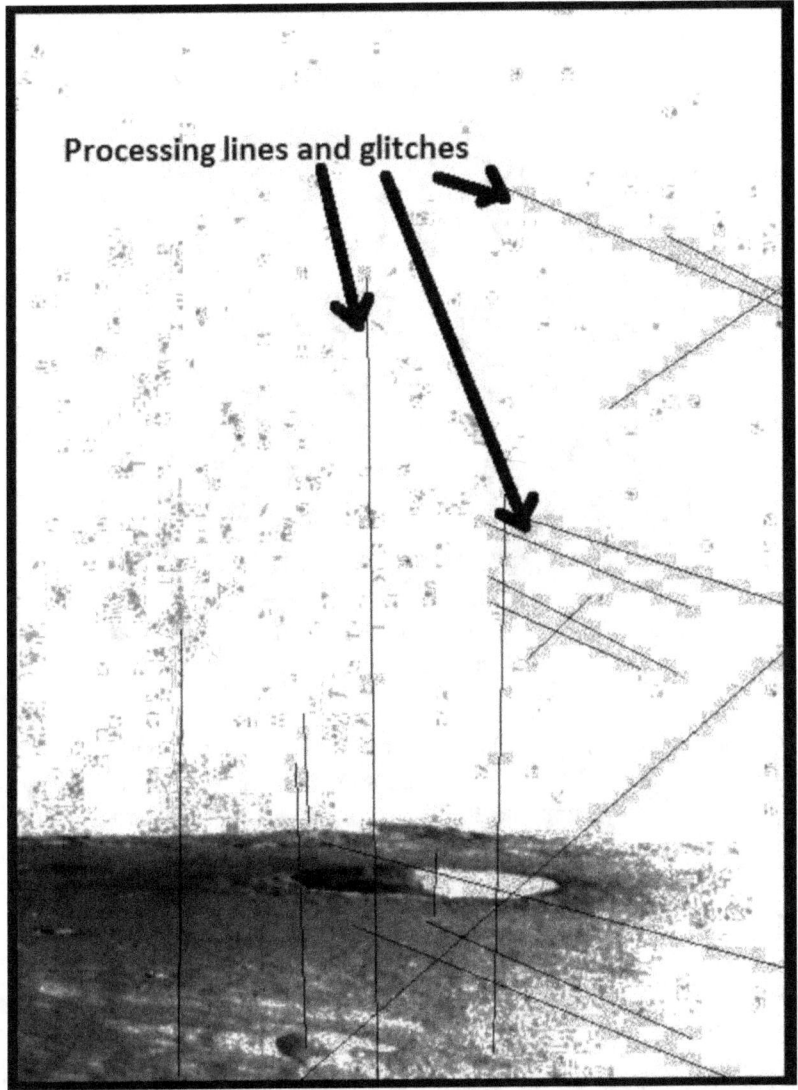

More scan and processing lines. Along these lines can be seen dangling debris, hairs and other blotches. (Negative Image)

14
THE "CASTLE" OF UKERT
(5.0 N x 5.0 E)

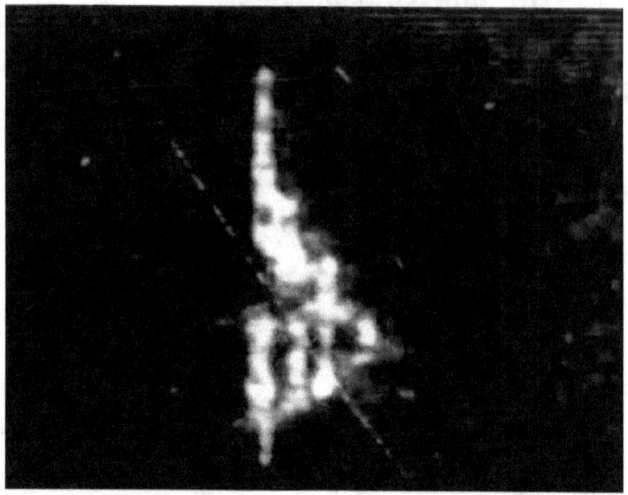

Hoagland's Famous dangling "Crystal Palace."

According to Mr Hoagland, there is much more evidence of an alien dome in the Sinus Medii area. Such so-called facts seem to be in total support of the concept that the Tower, the Cube, the Shard, and other atmospheric debris are but remaining fragments of a once far-larger, clearly artificial structure, once comprised of glass wired together and supported by some kind of darker structural framework. [23]

The Apollo-10 Mission shot a series of photos (Mag. S32) covering the Crater Ukert area North of Sinus Medii (A). The Photo AS10-32-4822 is one in the series covering this area. The series runs from frame 4809 to 4824. Photo 4822 is of special concern to Richard Hoagland. The remaining few "glass fragments," viewed in these photos are what caused Mr Hoagland to propose the one-time *"existence of an artificial, very ancient 'Sinus Medii Dome,' formerly completely enclosing this central region of the Moon."* The dangling glass crystal "Castle" is the last stop in Sinus Medii towards the final destination of the Mare Crisium dome complex.

In the video, "THE MOON-MARS CONNECTION" Richard reveals what he believes to be a great grid system or architectural structure of sorts, possibly indicating some kind of ancient lunar *"crystalline dome."* The dome lies somewhere around the Crater Ukert as he points out in the Ukert crater photo frame 4822. In this photo, scratches and all, he shows us "THE CASTLE," a small triangular shaped crystal structure dangling about in the air space, or vacuum space about 30-35 miles above the lunar surface! He locates it somewhere north and east of the crater Ukert.

The castle image of Mr Hoagland's MOON MARS Conference is located on a 20 x 24 inch b/w print, from the bottom right corner, at about 5 and 1/2 inches north by 3 and 3/4 inches to the left (west). Other photos that cover this area are AS10-32-4813, 4819, 4734 and L.O.III-M85.

Richard said, the "Castle" is the most important evidence (element) so far found, second to the Shard in significance of lunar geodetic dome construction. The crystal castle is one of the largest so-called *"bright slivers"* and *"glittering geometric fragments"* hanging in a supposed spider-web-like fragmentary framework of black invisible geodetic rebar support beams, high over the eastern end of Sinus Medii, estimated at 10-miles above the Lunar surface . We will soon see that 1000's of others seem to be hanging 110+ miles above the surface and along the surface of the photo.

Above, beyond and further towards the horizon lies Crater Manilius with its acclaimed "suspension-wires" and "trolley-cars." The most mysterious observable anomaly about his photo (AS10-32-4822) is the lighting. The Sun is at a mid-morning position, thus lighting all the left side brightness of the surface features. Nevertheless, the photo shows a fading or veiling of the lunar surface from left to right, where the right side becomes darker and more obscure.

The little Castle or crystal structure hangs brilliantly over the Mare area within this blurry obscure area, along with other glittery aerial objects, black rebar-like-beams and emulsion scratches. Mr Hoagland gives us his extraterrestrial explanation of these objects in light of his lunar dome theory. He said, *"With no definable 'lunar atmosphere' – rain, fog or clouds or any form of familiar terrestrial 'optical absorption mechanism' – the only logical explanation for this 'obscuration and fuzziness'* (after the possibility of simple photographic defects to the original 4822 negatives has been eliminated – which it has) *is that the Apollo astronauts actually photographed the remains of some kind of remarkable, constructed 'optical anomaly"* - stretching over Sinus Medii: A Semi-transparent glass-like, mechanical medium, with remarkably focused optical properties."

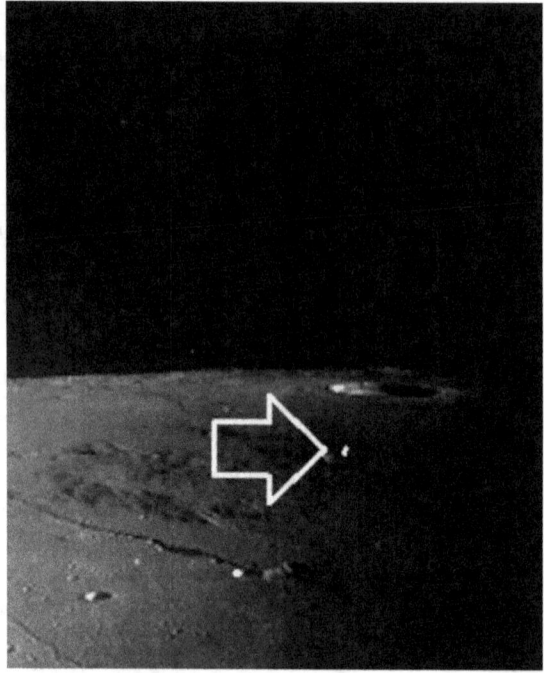

Lo cation of Crystal Palace on Frame 4822

Hoagland points out that this must be a fragmentary piece of glass-like material hanging in mid-air because it is being supported by an almost imperceptible wire-like slender cable, drooping and sagging under the weight of the object, very similar to the black suspension trolley-car wires about the Crater Manilius. Interestingly though, they are white rather than black.

It might be, supposes Richard that one type of wire reflects light, while the other does not. Nevertheless, both types seem to support the suspension of these crystal-like objects as testimony of a once fully functional geodetic dome.

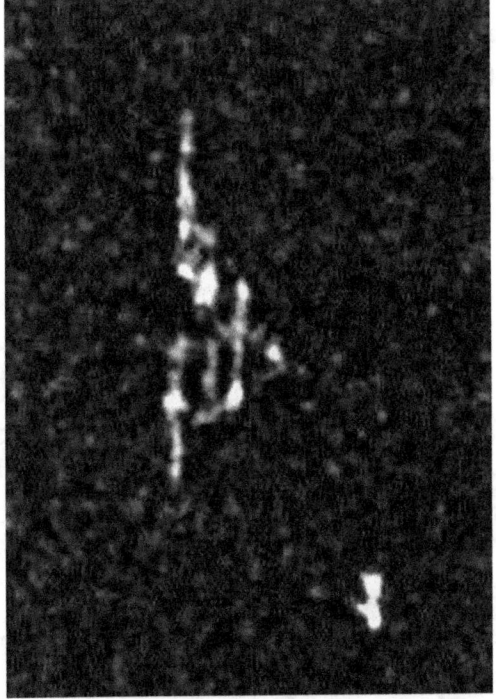

Closeup of processing glitch "Palace."

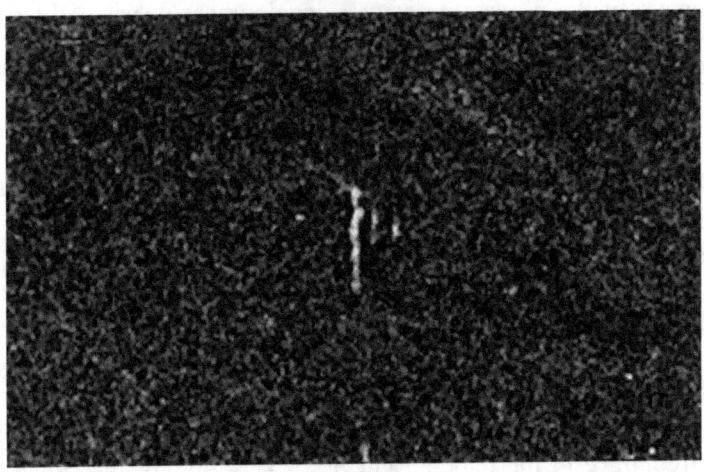

Hoagland's Famous dangling "Crystal Palace."

After reading all this super high technological babble and thumbing through reams of Lunar atlas photos, scanned, enlarged and enhanced images, it became necessary to order a NASA print of 4822 and check the fine details personally. The first contact at the NSSDC was the operator, whom kindly connected me with the librarian of the lunar image data collection.

After some flapping about the frame 4822, Mr Bob Tice replied that he was very familiar with that photograph, since Mr Hoagland and every other artifact hunter had been ordering copies for the same purpose. In fact, Mr Tice more than happily filled me in on a fiasco that took place in the film department between Richard and his associates concerning so-called alien artifacts. The Enterprise team had descended upon the image library collection like locusts, ransacked the files, culled out what they believed to be the best photo proofs for their theory and then split. They were out the door before the check settled on the secretary's desk!

The event climaxed with an argument between Richard and his assistants as to whether certain evidences were really factual or not. One of the associates was overheard saying to Mr Hoagland, *"Richard! That's absurd and ridiculous! You can't pawn that off with that interpretation on your reading audience."* i.e. even your most devout believers will not swallow that. I cannot be too sure what exactly it was that they were arguing over, and neither can Mr Tice, but it must have been relative to these crystal "things" or some other such similar nonsense.

Furthermore, the scene also exposed the fact that they were more than aware of the processing errors, development debris and scratches, and were concerned with just how far Mr Hoagland might go in fabricating alien artifacts out of glitches. Needless to say, after a few more conversations and laughs, the frame 4822 and other photo negatives were ordered, received and eventually studied.

After a few days of analysis, it became apparent that what was NOT mentioned in the MOON-MARS Conference tapes were the many hundreds of other "Castles" and crystal structures scattered all over the frames, especially frame 4822. One crystal of equal interest is located at about 8 inches from the right side of the photo and 5 inches northward.

This dangling derivative also displays the SAME brilliance and reflectivity as the Hoagland Crystal. Another interesting feature is both have similar highlights and shadows cast in the same direction. In the photo (contrary to the darker right half of the photo) the sun is to the right and is casting shadows from the surface features to the left.

The dangling object displays the same shadow and highlight phenomenon. By imaging and enhancing these and other surrounding areas, lines can be seen crisscrossing north and south, east and west like a grid system with some running diagonally and obliquely in different directions. Other 'scratches' can be found in many locations in the photo, from 75-90 miles high to 110 miles high if one calculates according to Hoagland's geometry.

Are these pieces of a huge crystal dome structure or emulsion scratches and debris in the development of the negatives? Every scratch and emulsion pattern flaw displays this same castle crystal-like structure to a greater or lesser degree. The following photos demonstrate the wide variety and placement of such errors as well as prove that the crystal-like dangling castle is not unique.

The dangling crystals and hanging shards of glass are found everywhere in and on the photo 4822. They are found in the middle, on the sides, scattered throughout space or what appears to be the surface of the photo. It seems that if the photo could be enlarged and expanded beyond the lunar image area the little crystals would also continue as well.

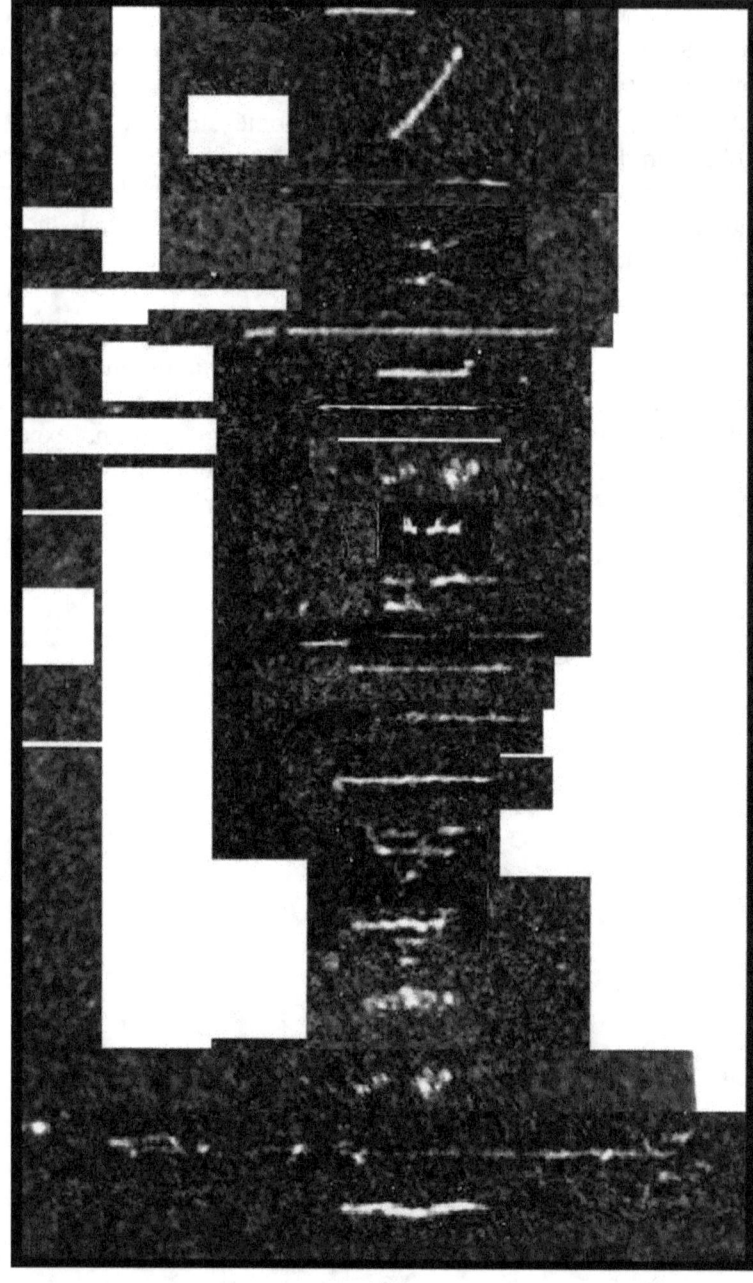

(A)
Compilation collection (copy, cut and paste) of crystal-like structures taken from multiple areas of frame 4822

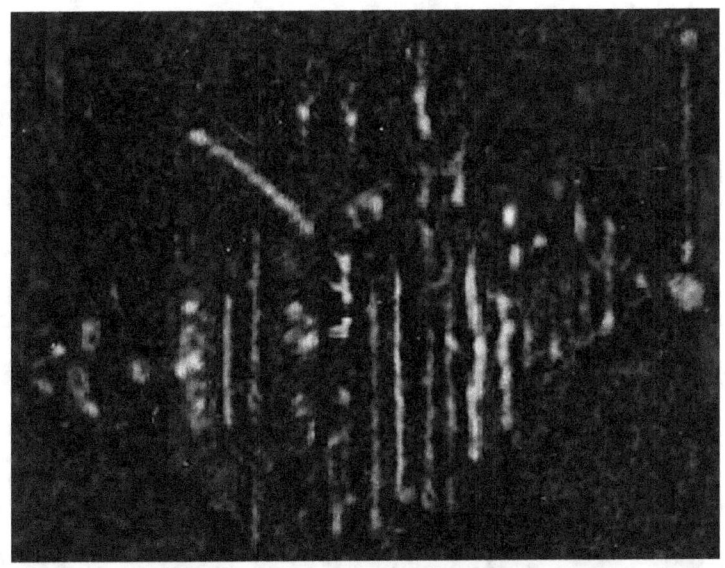

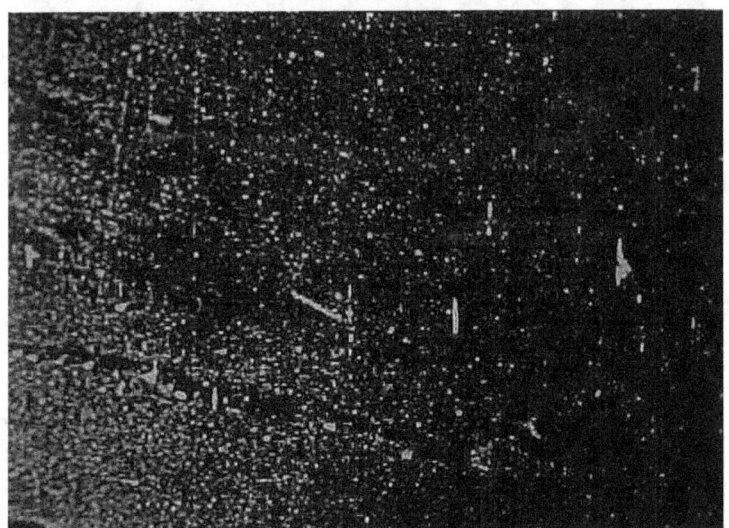

(B) **Compilation of other crystal-like objects.**

The following demonstration is the best explanation for the dangling crystal phenomenon. If we split the photo into two layers [See "A" and "B" above], the bottom layer "A" [demonstrated with negative image] would be the clean image of the lunar surface area around Ukert. The top plain "B" would be the surface area of the printed photograph with the development artifact inclusions. Notice that the scan lines and the so-called alien crystal-like structures run perfectly level and flat with the surface of the photo and not within the photo as if hanging "above" the lunar surface.

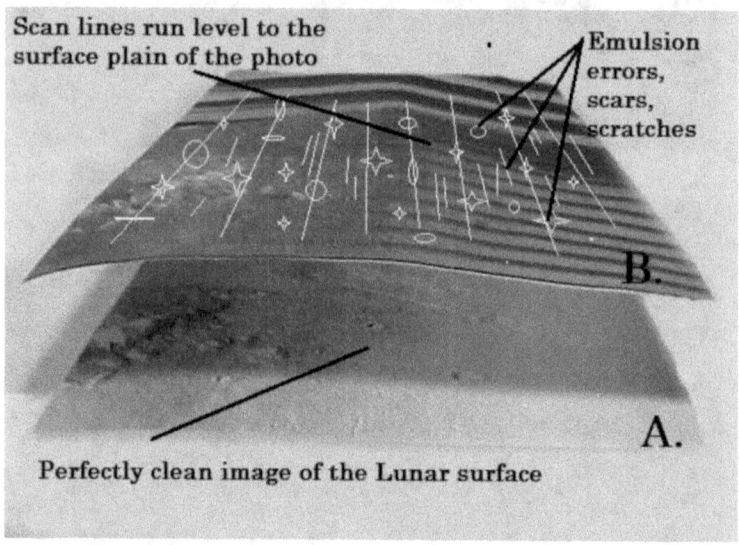

The above is the best explanation for the dangling crystal phenomenon. If we split the photo into two layers [See "A" and "B" above], the bottom layer "A" [demonstrated with negative image] would be the clean image of the lunar surface area around Ukert. The top plain "B" would be the surface area of the printed photograph with the development artifact inclusions. Notice that the scan lines and the so-called alien crystal-like structures run perfectly level and flat with the surface of the photo and not within the photo as if hanging "above" the lunar surface.

Therefore, it is now proved that the dangling crystal-like structures are not alien artifacts hanging above the lunar surface within ancient decayed structural support beams but are developmental artifacts embedded within the surface emulsion of the photograph. Furthermore, once these artifacts are stripped away and the observer is able to see clearly that there are no towers, support beams, fragmentary glass crystals, abandoned trolley cars, streets, highways, bus stops or any other alien manufactured device. The Lunar surface is finally exposed as sterile of all such alien nonsense.

15
THE MARE CRISIUM DOME AND CRYSTAL SPIRE
(AS16-121-19438)

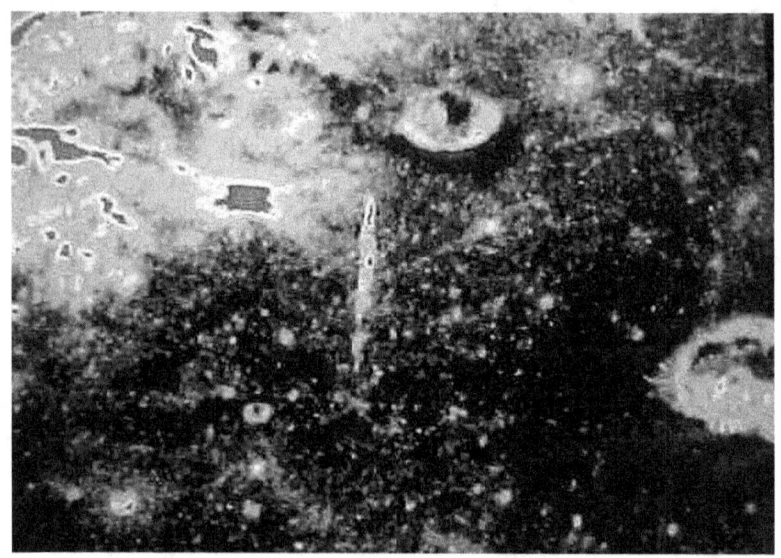

NSSDC PRINT of AS16-121-19438

The Sea or Crisis [Mare Crisium] is another location Hoagland leads us to in searching for massive alien constructions. Hoagland shows us an especially unique area, in fact the only photo example [AS16-121-19438] of a set of large circular light patterns within the Mare. If one bends their visual senses under extreme eye strain, as well as their rational senses, they will see what appears to be an enormous "spire" or tower rising from the lunar surface into the sky within one of these circular light patterns.

When enlarged to an outrageous proportion this "spire" tower demonstrates the same layered "pixel-like" crystalline and cubic patterns as seen before in the Tower and Cube. The concentric circles of light over the surface are considered by the artifact hunters to be light reflection and refraction glistening through the remaining portions of the glass dome. The spire they say [25] is a 20 mile high version of the same crystal-like glass structure "shard" located directly west of Crater Picard.

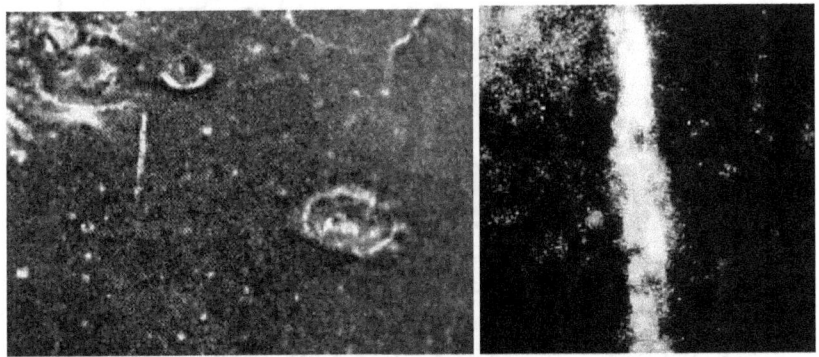

Close up of Spire showing the refractive properties of the so-called crystalline glass structures of what is left of a tower according to Richard Hoagland

Location of the "Spire" LO-IV-191-H3

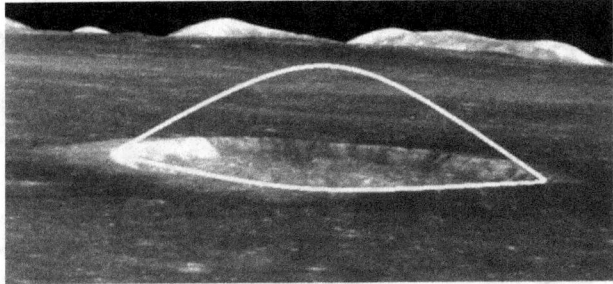

Picard crater and supposed dome (outlined in white) from Apollo frame AS10-30-4421

Telescope photograph

Interestingly the spire cannot be found on any other NASA Lunar photograph no matter what angle or time of day the photos were taken. Comparison of the NASA print (AS16-121-19438) with earth based telescopic photographs and copies of NSSDC versions revealed several remarkable discrepancies. One extraordinary feature on the original NASA print turned up completely missing: a brilliantly illuminated spire, at least 20 miles in height, located directly west of crater Picard. [26]

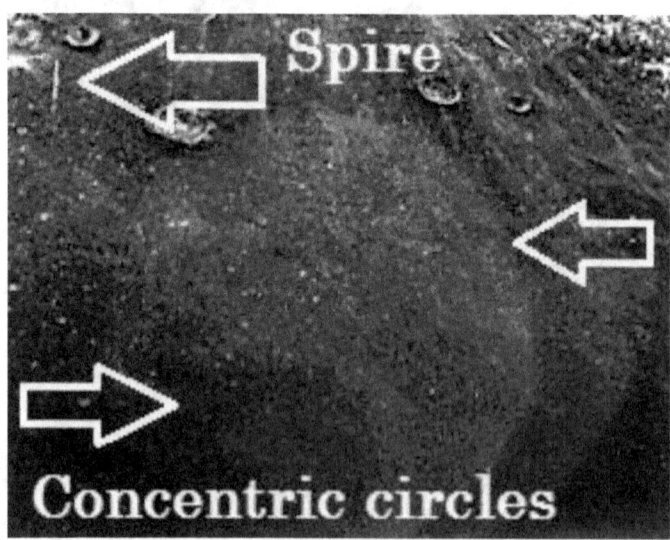

Possible larger "dome" over Mare Crisium

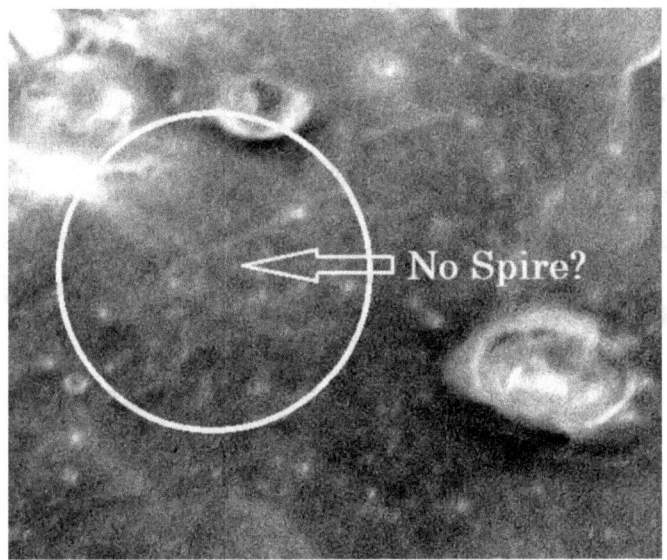

**SPIRE is missing on the official NASA version of
AS16-121-19438**

Hoagland blurts out it is missing on the official NASA print (the original) in his Martian Horizons and exclaims, "*I kid you not!*" [See his figure 39 on page 51]. Of course he is not kidding us. It is a fact that only on the NASA copy at the NSSDC can the "spire" be seen. His excuse is that (choke!) the light has to be just right to refract and this has to happen just at the right time the astronaut is snapping the picture. Otherwise, the high allusiveness of the crystalline material makes it almost impossible to see.

Notice though, the same location shot by Apollo-11 of Crater Picard (AS11-42-6223). In this shot the Sun is to the right side causing shadows on the right insides of the local craters, while brightly lighting the left insides crater walls. Crater Picard is in the middle of this shot with no indication of any "spire" or 20 mile high tower. The spire should be seen in this shot sitting 20 miles southward in the frame, since the camera is much further away than in the other shot taken by Apollo-16.

None of the Lunar Orbiter Missions photographed the spire (LO-IV-191-H3). Apollo-11 did not capture the spire (AS11-42-6220-6223 and 6214).

Apollo-10 did not snap any picture of this spire either (AS10-30-4417, 4419 and 4421). Neither did Apollo-15 (AS15-M-0954) or the latest photography taken by the Lunar Orbiter Reconnaissance Camera

(SPIRE and Picard Crater LROC QuickMap). The real crisis in the Mare is that there are no glass domes or spires and this is very dis-spiring to those who want aliens to be found on the Moon.

AS11-42-6214

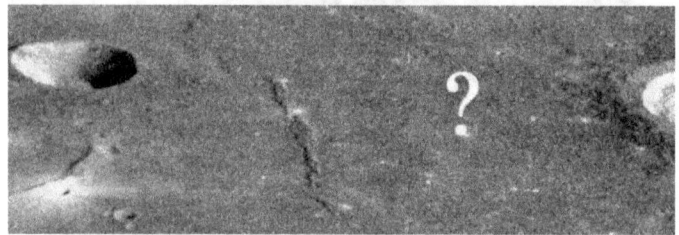
AS11-42-6223

AS11-42-6221

AS11-42-6220

AS10-30-4417

AS10-30-4419

AS10-30-4421

AS15-M-0954

LROC QuickMap

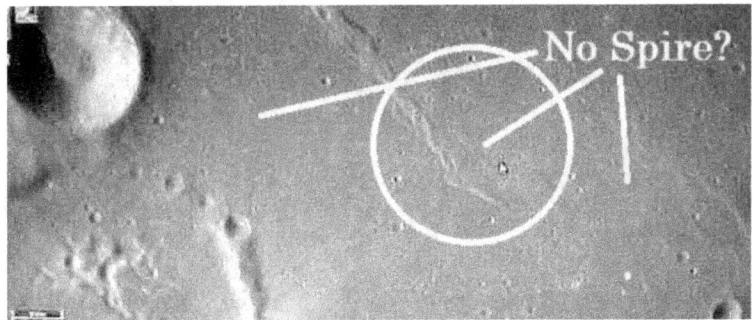

SPIRE missing near Picard Crater (LROC QuickMap)
Lat. 13.0 x Long. 54.0

Nevertheless, no matter what arguments are raised about the validity of the dome and the crystal spire, Richard continues to propound that on a closer examination of this spire through photographic enlargement and enhancement, it becomes readily apparent that it strikingly resembles a highly damaged cylindrical shaped tower frosted with glass and complete with "fins" as well! [26]

Well, after reading this cock-and-bull statement, which seemed fishy to me, I studied the photo myself and found, after "photographic enlargement and enhancement," that the "spire" in question rather strikingly resembled a huge, ugly and irritating development scratch.

While referencing the original NASA print with the NSSDC copy along with all the other photographs of the same area, it became "readily apparent" why the original print of 19438 and all the other photos of the area were missing the "spire," while the Greenbelt Maryland copy contained it.

This most "remarkable discrepancy" as Richard called it was no more remarkable than any other photo development artifact or film defect. The reason why the original lacked it while the NSSDC copy had it was that the copy got scratched somehow by mishandling in the process of usage. The other interesting thing is that this particular photo is one of the best, if not the best, shot of the area and was chosen for publicity purposes. Naturally, the scratch had to be inked over and corrected. Hence the reason why in many P.R. Photos the scratch is seen retouched and blotted out for visual purposes. Why the NSSDC did not obtain another clean copy is unknown. Maybe it was easier to just retouch what they had. I am sure Richard could make some conspiracy out of this if he has not already done so.

16
MORE UKERT CRYSTALS

Another area of alien interest for dumb dome seekers is the area northeast of Crater Ukert. It is a part of the hill behind (north of) the linear dome-shaped hill, which runs diagonally across the photo. It ends to the right at a small fault line or canal. In the foreground of this hill at Ukert crater, is a small square or diamond shaped field. Maybe an Alien baseball field for small alien children to play in? Or maybe foundations to walls that once existed? These two areas display unusual features suggestive of unnatural structures and surface anomalies.

The large area north of the diagonal hill contains regularly aligned rows of structures according to Dr. Bruce Cornet. There are more than a dozen small craters peppering the area, modifying the lunarscape or cityscape. At a distance, the rows appear like benches, platforms and building complexes, said Dr. Cornet.

Dr. Cornet suggests that, "*on earth, such a feature would be interpreted as the pattern produced by the eroded edges of layered rocks that dip below the surface*". Yet he argues, that "*on the Moon, there have been no physical processes that can account for such a regular geological structure (for) rilles and wrinkles on the surface of a cooling magma outflow do not form such a regular pattern as is evidenced in so many Mare areas on the Moon.*" His most interesting geological pattern point being, that this anomalous pattern has definite boundaries by and which it is absent. [24]

Upon computer enhancement and magnification, the pattern takes on other characteristics. Rectangular features show up along the benches or rows, with some having gaps between them. Thin spires project up from the surface in several places along the rows. Some of the rectangular structures take on a form like buildings and high rise apartment buildings.

The whole area according to Mr Cornet, resembles what one might expect for a city the size of Los Angeles, that has been abandoned by its occupants and left behind, maybe after an atomic bomb drop. It looks like a decayed city complex! Too geometrically laid out to be natural and too artificial to be natural lunar surface features!

Surface processes such as meteoric bombardment have provided the abrasive forces that have destroyed and obscured much of the details. Yet, the ghostly image of an ancient alien, lunar-domed cityscape appears to stand out among the natural surface features and craters that surround Crater Ukert.

But, are these the ravings of an alien artifact hunter and the pseudo-scientific dribble of a retired, bored, burned out geologist? Does Richard Hoagland and persons like Dr. Cornet see things that are truly evident to alien occupation in some far distant past? Only the further imaging of these and other photos will tell us the truth, the whole truth and nothing but the truth. But, don't waste thousands of dollars on NASA negatives to find this truth out. Just look at these examples and save your hard earned money. I wish I had!

Let us look at some shots of the City in finer detail in these imaged photo enhancements. You will notice that there are many crystals, more than expected, ALL running along a photo development path, all causing the same emulsion pattern. Look at them ALL together and compare them. The whole photo from every square inch has these glitches and scratches. All run along a line contrary to some rounded dome shaped pattern, as suggested by Richard Hoagland. As to the City Complex and Dr. Cornet's theory of some abandoned dwelling place, there are many lunar surface areas with the same potential for alien city complexes. The whole surface of the moon may be said to have been occupied at one time.

Only a LUNATIC with a Mickey Mouse imagination would propose such a thing as these guys have. According to previous studies, Richard Hoagland and his followers do the same hoaxing as George Leonard and Fred Steckling! It has been said that Richard has been called out many times by his own technicians, that he was seeing things that should not be passed off on to the public: *"Richard, you should be ashamed of yourself, to think that others will see what you see! You cannot pass these interpretations off on to the public with a clear conscience!"* [Quote: A NASA employee who overheard them talking at the NSSDC photo library].

But, needless to say, Richard's conscience is as clear as his crystal glass domes he talks about, when it comes to pawning off artificial-looking cities, domed craters and dangling crystal-like glass trolley-car cable systems. I expect, as age creeps up on our fellow mythologist that we should expect seeing from him a blurry picture of a McDonald's in some crater (27).

17
NASA COVERUP OR TOUCHUP?
A Study of Lunar Shadow, Highlight, Light Reflection, Refraction, and Processing Anomalies:
(AS10-32-4734)

Supposed NASA cover up (inking-out) of alien crater habitations

One photo that mysteriously shows up supposedly censored by NASA (according to the Hoaglander's) is AS10-32-4734 (Apollo 10 Mission, Mag. 32., Photo Number/Frame 4734). The photo can be viewed in the NASA publication SP-246, "Lunar Photographs From Apollo 8, 10 and 11" [28]

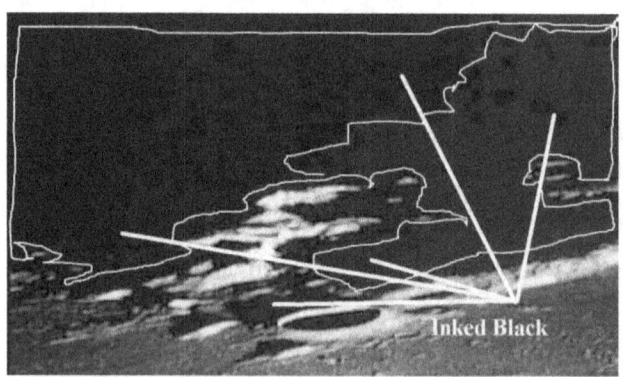

Logetronically processed retouched "inked" photograph

The photo is a westward shot looking past Crater Triesnecker over and beyond Crater Chladni, further over the Murchison and Pallas-E areas, just south of Crater Pallas. The area location under discussion is 2.5 degrees North to 6.2 degrees North, by 2 degrees West to 2 degrees east.

The area includes part of Crater Pallas, Crater Chladni and can be seen in part in Lunar Orbiter Mission III, Medium Resolution Photo Number 85. The area is on the southern tip of the highlands that extend from the North into Sinus Medii. [29]

The obvious problem with this oblique (orbital) shot, in this photo, is that it has a lighter "gray" tone quality. Yet, mysteriously it has darker shadows on much of the lunar background landscape, unlike some of the other lighter gray shadows that lay around the lunar mare area. Remembering that the Lunar surface does not lend itself well to light refraction [almost none at all], because of the lack of atmosphere, it stands apparent, that the differences between the shadow's tones is an anomaly or an artist's touchup. It may be some kind of development process. All shadows should have (according to other scientists) the same-pitch black look with no light refraction.

In this photo, the two different shadow tones betray censorship, the conspirators say. No shadow should be dark black while others are light gray.

Observe photo 4734 shot of Crater Triesnecker. The crater's shadow to the left is lighter gray. The inside crater shadow that lies along the crater floor is extremely dark black, like India Ink! Also, the shadow on the crater's floor conforms to the parameter of the floor rather than to the edge of the rim casting the shadow! (Somebody goofed?)

Next, take a look at the supposedly natural looking density of the shadow areas beyond Crater Chladni and compare them to the shadows about the Mare floor among the shadow areas of the various small ridges. This photo has obviously been "doctored" with jet opaque, black, retouch ink and some kind of retouch process has been performed on it. The ink job covers all the area from the horizon (which is in the Moon's terminator?), to the edges of the highland area where the Sinus Medii Mare area begins. One of the areas left out is the West Side shadow of Triesnecker. Has some NASA mole fixed up this particular photo? Has something been hidden from our view?

Now, the following questions arise. Why would NASA cover up half this photo with re-touch black ink? What would they be covering up? Are they covering up something? If they are, why publish the photo at all? If they are not, why retouch the photo, when the image could stand on its on? My sources tell me that the original negative does not have this retouching job on it. So, what is the reason for retouching or fixing up photo 4734, when all it would be showing is common lunar highland geography? For, the same area can be seen in LO-III-85 medium shot. [30]

Photo M85 is a Lunar Orbiter-III shot of the same Crater Murchison with Crater Ukert in the distant background. The area covered here, which is plainly visible, is the area "inked-out" in AS10-32-4734. Why black out one and not both? Maybe because 4734 is an oblique view looking across at a lower angle, while M85 is from a much higher angle, and 4734 would reveal allot more than M85? My god! how far should one take conspiracies?

My sources also tell me, since I do not have a copy of 4734 at the moment, that the untouched 4734 photographic negative depicts a good quality landscape shot of the surrounding Murchison area, which extends into the lunar terminator (dark-side). Could there be something more visible in the 4734 photo that is not seen in M85? But some will justly suggest that 4734 was touched-up for some purpose unique to the publication of SP-246. Yet, no good reasons can be given for inking-out surface features in a photo for this publication, short of hiding something! The photos and their shadows themselves should more than suffice in standing out on their own.

LO-III-M85

The SP-246 publication has other photos of less quality than 4734, and they seem to not be tampered with? Why not ink them or logetronically process them too? Why only 4734? Why retouch any photo? Is this not destroying research potential? Now! Have these NASA moles and conspirators really censored anything?

The Hoagland hoax theorists say, chances are, some others have been inked in, where they deal with Crater Ukert and Triesnecker, and the areas surrounding Northern Sinus Medii. Is something in the Murchison Crater area being purposely obscured by NASA? Are they hiding something?

Further studies and calling turned up more on this subject. I spoke to another friend of Bob Tice; a NASA photo-technician named Jay Freidlander, about this suspected fraud or re-touching. According to a conspiratorial mind, the arguments would sound somewhat suspiciously like a cover-up or something, but hearing the guy out led me differently.

His response actually answered all my questions. He said, that if you look at some of the other photos in the publication, SP-246, such as with AS8-13-2347 [on page 19], and AS8-13-2314 [on page-17] you could see the same similar effects. This effect has nothing to do with any cover-ups, as suggested by Hoagland, William Brian, Fred Steckling or George Leonard. The hiding of lunar alien bases, ancient cities, glass domed structures, crystal palaces or some other imaginative things just cannot be proved!

A large portion of the Apollo photos, they say, were processed two different ways: Photometrically and Logetronically. Photometrically, the photos retain their original look, as taken by the Hasselblad Camera. Logetronically developed, photos undergo image gain adjustments, where both the highlights and shadows are enhanced by high and low gain adjusting. High Gamma gain brings out the details in the highlight areas, whereas Low Gamma gain reproduction, the normal procedure, retains the original details, that is, if the photo is bad, the details are lost or are not very strong in details. Logetronic processed photos amplify where the result is the highlights are brighter and the shadows are darker: *"Different types of reproductions were made of the films. For Apollo-15 film, a logetronic process was used in which the reproductions were made with quality control on the density using exposure control and dodging techniques. A photometric procedure using a Niagara printer that allows photometric and photogrammetric (albedo) measures to be made processed Apollo-16 film and a second set of Apollo-16 film. The process used for Apollo-17 film involved gain adjustments, and both high gain and low gain reproductions were made. High gamma (gain) brings out details in the highlight areas. Low gamma reproduction is the normal procedure, but details in highlight and shadowed areas are lost.* [31]

This process was used in the early periods of photography during the Apollo Programs, for enhancing, before Computer enhancement began. The complete photos can be obtained either in Photometric or Logotronic processed formats. Early Logotronic enhancing is similar to modern computer line-art imaging. The Logetronic process was discontinued by NASA for obvious new technological reasons, as well as scientific and public complaints.

My contact also said the Apollo AS10-32-4734 photo in the SP-246 publication is a Logetronic processed photo and can also be purchased as a photometric, which will contain all the details and lack of details in its original form, which in some cases the logetronic version obscures. He said, there are absolutely NO conspiracies going on inside the NSSDC in hiding or altering photos, either to add or obscure any features. Just purchase the photometric versions and you will be studying photographs for their lunar geology and surface features.

I crossed referenced this location, which covers the area of Chladni, Murchison and Ukert, Craters located at the southern tip of the highlands extending into Sinus Medii, with other photos. The top portion extends into the dark area of Murchison. But is NASA supposedly hiding something here behind the logetronic darkening process? The darker area is covered in detail in the Lunar Orbiter photos LO.W-H-102 and 109. These cover the same area as seen in LO.III-M85, which is a top shot and is as clear as a bell in the Lunar Orbiter Photo album SP-200 NASA publication, page 56. [32] Therefore, no conspiracy to hid any weird or strange alien archaeology is evidenced here.

Later, as I studied the SP-246 photo album, I ran across a color photo with the same apparent 'logetronic' look, photo AS11A0-5921, a Hasselblad photo of the under-side of the Lunar Module, which only shows the exhaust bell and the blown away surface regolith under it. What is the problem? The problem is, only B&W photos were logetronically processed and this one is a color one!

One will notice the same crater shadow density different here in the rock in the lower left-hand corner. The rock's shadow is lighter than the shadows cast by the LEM's legs and exhaust bell. Also, looking up in the top left, at the equipment box (?) next to the other leg, the shadow of the object is gray, while the shadow is pitch black? Could the color photos also have been processed logetronically? Maybe the later technicians have forgotten? Or is this some kind of 'touch-up' work done with opaque black ink to enhance the photo for public consumer publication consumption in the SP-246 NASA publication?

Mr Freidlander said, no to both. But, the shadows are different and are of different densities. The LM's leg shadows are darker while tile rock's shadow and the equipment package are gray. One suggestion Freidlander said is that the process of printing the photo in the SP-246, in 1971 is, of course, a dot matrix printing. Tiny dots make up the color or B&W picture for mass printing. It is not a formal photographically developed picture. Order the photographs and compare them. [33]

Shadow and highlight play on the lunar surface is an oddity to the untrained and has its unique features and effects. Selenographers say there is a difference between refracted light and reflected light. Light does not refract very much at all in the lunar atmosphere, which is very thin. Light reflection is different. Light travels no matter what, and if it has a surface it hits against, it will light that surface up.

An example is the Lunar Module's gold colored reflection material for decreasing and reflecting the heat off the spacecraft. This is highly reflective material, whereas the rocks are flat with no shinny reflective surfaces. Hence, the rocks do not reflect much light back into their shadows, let alone refract any light, while the Lunar Module gets its whole dark side lit up from reflection.

Other higher reflective material surfaces are the manmade space suits, equipment, and others electrical devices used on the Moon's surface. All the organic materials, such as the rocks and minerals and surface dust are flat grays and are not very good reflectors of light.

I checked through my Apollo kinescope footage of Apollo-17. In one TV Camera shot as the astronauts walked up to the big rock and when they entered the area where the shadow was, the shaded side of the rock began to appear. There was some reflected light bouncing off the dark shadowed side, where the astronauts stood. When they backed off and as the Camera changed distance, the shadow darkened to pitch black again. Yes, there is even some minor reflected light going into the shadows of the lunar surface features. [34]

Yes, there may be some few touch-up jobs here and there, as any lab and distribution company would do. And yes, there may be some simulations or film editing for some time and cost factors, but almost every photo shop does touch-up work on people's faces, as they have done the craters. NASA may have touched up a few photos for publicity sake.

Every organization does this to impress the people. But, when one looks through all the 1000's of Apollo shots, they will see that a small percentage have that perfect studio look. Most are bleached out because of bright sunlight, or too dark for a lack of it. The astronauts were not your top notch photographers, and even if they had been, they were under such a heavy time restriction that, even a professional would have made mistakes.

As far as the touch-up work, the technicians in the lab have asked me, *"Would you like us to clean up some of the dirty photos, as the masters are getting old and scratched up?"* I said no, of course, for I preferred all the glitters, scratches and potentially "hidden" things.

Yet, overall, the NSSDC does not touch up, nor do they restrict you from purchasing any particular photo, as I have dealt with them personally and have ordered many lunar photos, films and video tapes. Actually, anything I request that I want, they will sell a duplicate of it.

No, I don't believe that there are glass domed ruined cities on the moon nor any mile high dirt digging machines, sucking machines or blowing machines lying around. As far as the Apollo missions, there are plenty of research materials available and I have seen plenty.

I cannot find any solid arguments of hoaxing or censoring or hiding going on. The NSSDC answered every question I had. Remember, NASA and the NSSDC may be "GOVERNMENT" facilities, but it ALL belongs to and is accessible to U.S. Citizens and professional scientists alike, as long as the small fees are paid for the processing of the data.

- FIN -

FOOTNOTES

SECTION-3
FRED STECKLING
FOOTNOTES

[1] See, 44,52,54,58,60,70,72,74,76,78, 80, 82, 84, 86, 88, 90, 92, 94, 98, 100, 102, 104, 106, 108, 112, 114, 120, 122, 124, 125, 126, 127, 128, 130, 136, 137, 138, 139, 140, 143, 144, 145; (151 is Fred's Pond Photo), and 154, 155, 156, 157-3, 158-3, 160-3, 169,183, 178-1.3, 178-1.4, 185-1.3, 185-1.4, 192-1.4

[2] See, 55-3, 82-1, 82-2, 86-3, 90-1, 91-1, 92-1, 92-3, 103-1, 121-1, 148-2, 149-1,2,3; 163-2,3; 164-1, 168-3, 182-1,2.

SECTION-6
RICHARD HOAGLAND
FOOTNOTES:

[1a]. Dr. David Morrison, Chief of the Space Science Division at NASA's Ames Research Centre, that Hoagland was largely "self-educated" in science. In an August 31, 1990, letter, Morrison told me that he knew of "no one in the scientific community, or who is associated with the NASA Mars Science Working Group, or who is working on Mars mission plans at such NASA centers as Ames, Johnson, or JPL, who ascribes even the smallest credibility to Hoagland or his weird ideas about Mars."

[1] "Cydonia - the face on Mars." ESA. September 21, 2006. http://en.wikipedia.org/wiki/Mesa

[2] "Planetary Names: Mars." Gazetteer of Planetary Nomenclature. USGS Astrogeology Research Program. And, "Planetary Names: Feature Types." Gazetteer of Planetary Nomenclature. USGS Astrogeology Research Program. http://en.wikipedia.org/wiki/Mountain

[3] Hoagland's Martian Horizons (M.H.) V2:5, p.3

[4] "We Found Alien Bases on the Moon" Fred Steckling.

[5] "Somebody Else Is on the Moon" George Leonard.

[6] M.H. V2:5, p.4

[7] http://www.lpi.usra.edu/resources/apollo/catalog/70mm/magazine/?32

[8] M.H. V2:5, p.29

[9] M.H. V2:5, p.15

[10] "Surveyor Observations of Lunar Horizon-Glow." By J.J. Rennilson and D.R. Criswell, The Moon, Volume 10, Issue 2, pp.121-142. 06/1974.

And,

http://www.thelivingmoon.com/43ancients/02files/Surveyor_06.html

And, [http://en.wikipedia.org/wiki/Surveyor_6]

[11]

http://www.thelivingmoon.com/43ancients/02files/Surveyor_07.html

[12] NASA Technical report 32-12162, Surveyor 6, Mission Report, part-III, JPL. Aug. 15, 1968

[13] Bruce Cornet. Misc. Correspondence, photo notes No. 16, 17 and 18. Interpretations, Correspondences to Richard Hoagland: April 24th, 28th, May 11, and July 4th 1994.

[14] The Moon has an atmosphere, but it is very tenuous. The Lunar Atmospheric Composition Experiment was deployed on Apollo 17. It was a mass spectrometer that measured the composition of the lunar atmosphere. On earlier missions, only the total abundance of the lunar atmosphere was measured by the Cold Cathode Gauge. The three primary gases in the lunar atmosphere are neon, helium, and hydrogen, in roughly equal amounts. Small amounts of methane, carbon dioxide, ammonia, and water were also detected. In addition, argon-40 was detected, and its abundance increased at times of high

seismic activity. Argon-40 is produced by the radioactive decay of potassium-40 in the lunar interior, and the seismic activity may have allowed escape of argon from the interior to the surface along newly created fractures.

http://www.lpi.usra.edu/lunar/missions/apollo/apollo_17/experiments/lace/

[15] Moon Dust: Every Apollo astronaut did it. They couldn't touch their noses to the lunar surface. But, after every moonwalk (or "EVA"), they would tramp the stuff back inside the lander. Moon-dust was incredibly clingy, sticking to boots, gloves and other exposed surfaces. No matter how hard they tried to brush their suits before re-entering the cabin, some dust (and sometimes a lot of dust) made its way inside. Once their helmets and gloves were off, the astronauts could feel, smell and even taste the moon. What is moon-dust made of? Almost half is silicon dioxide glass created by meteoroids hitting the moon. These impacts, which have been going on for billions of years, fuse topsoil into glass and shatter the same into tiny pieces. Moon-dust is also rich in iron,

calcium and magnesium bound up in minerals such as olivine and pyroxene.

http://science1.nasa.gov/science-news/science-at-nasa/2006/30jan_smellofmoondust/

"First and foremost is just the fact that the dust just sticks to everything," said Jasper Halekas, a research physicist at University of California, Berkeley Space Sciences Laboratory in Berkeley, California. From gauge dials, helmet sun shades to spacesuits and tools, the "stick-to-itness" of dust during the Apollo missions proved to be a noteworthy problem, Halekas reported. Most amusingly, he added, even the vacuum cleaner that was designed to clean off the dust clogged down and jammed. Halekas recounted a technical debrief by Apollo 17's Gene Cernan after his 1972 Moon voyage. Cernan said that "one of the most aggravating, restricting facets of lunar surface exploration is the dust and its adherence to everything no matter what kind ... and its restrictive friction-like action to everything it gets on." The astronaut added: "You have to live with it but you're continually fighting the dust problem both outside and inside the spacecraft."

Although the lunar environment is often considered to be essentially static, Halekas and his fellow researchers reported at the workshop that, in fact, it is very electrically active. The surface of the Moon charges in response to currents incident on its surface, and is exposed to a variety of different charging environments during its orbit around the Earth. Those charging currents span several orders of magnitude, he said. Dust adhesion is likely increased by the angular barbed shapes of lunar dust, found to quickly and effectively coat all surfaces it comes into contact with. Additionally, that clinging is possibly due to electrostatic charging, Halekas explained. "I think it would behoove us to understand the lunar dust plasma environment as well as possible before we try to come up with detailed dust mitigation strategies," Halekas told SPACE.com. "This would mean characterizing the dust, electric fields and plasma around the Moon and understanding how they interact."

http://www.space.com/3080-lunar-explorers-face-moon-dust-dilemma.html

[16] [http://www.windows2universe.org/earth/moon/lunar_atm.html]

[17] [http://en.wikipedia.org/wiki/Rayleigh_scattering]

[18] LO-III-84M; LO-IV-97M, 101M, 101-H3, 102-H1, 102M, 108-H3, 109-H1 and 112-M

[19] AS16-121-19438, AS10-32-4854, 4855, 4856 and the previously discussed frame 4822.

[20] See Lalande photos: Virtual Moon Atlas, LOPAM_Lalande, LAC_LM, Lalande_LM77, Apollo Mapping Camera Lalande_A16 2. And, LO-IV-114-H1, 113-H3, 108-H3 and 109-H1.

[21] http://wms.lroc.asu.edu/lroc

[22] M. Hor. p.43 and 44

[23] M. Hor. p.22

[24] Cornet, Dr. Bruce: (Geologist) Handout notes published through NYWK SVCS Systems and MARS MISSION. 5/16/1994, p.6. [See PHOTO REFERENCES:

AS10-32-4734, AS10-32-4813, AS10-32-4819, 4809 and 4824. Also, L.O.IV.85M]

[25] Martian Horizons p. 51

[26] Ibid. p.51

[27]   http://tomcornett.hubpages.com/hub/Apollo-Moon-landing--oh-really

[28] Lunar Photographs from Apollo. 8, 10 and 11. NASA SP-246. Printed by the

U.S. Govemment Printing Office. Washington, D.C. 1971. p.68

[29] U.S. Geological Survey Lunar Map #1511, Mare Vaporum Quadrangle, US Geological Survey, Denver, Co. 1968. (Northern Sinus Medii "Pallas" area).

[30] The Moon as Viewed by Lunar Orbiter. NASA SP-200. 1970 ed. by L.J.Fosfsky and Farouk El-Baz. p.56.

[31] NSSDC WDC-A-R&S Catalogue of Lunar Mission Data, July 1977, Page 85.

[32] Compare photos AS1042-4734, L.O.IV-H-102, 109 and L.O.III-M85.

[33] Phone conversation between and Bob Tice and Jay Freidlander of the NSSDC, August 17, 1995

[34] Apollo-17 TV Kinescope transmitted footage (Tape-2) of astronaut's suit, ALSEP Site Setup 347-03-[12] to (39), time code at 0.17.00m

and Tape video 3-4.

See, Photos and films studied and compared:

AS10-32-4734

L.0.IV-H-102 and H-109

L.0.III-M85

As8-13-2347
As8-13-2314
As11-40-5921
PHOTO REFERENCE OF MARE CRISIUM
N.W. Mare Crisium / USGS I-799
LO-IV-H-61, and H-66
Apollo-15 frames 9237-9230 and 9494-9490
S.W., S., S.E.M. Crisium, USGS I-837 and I-948
LO-IV 61-H, 54-H, 177-H, 191-H, 192-H, 184-H and 185-H, 191-192-H (Low Sun angle).
Apollo-=10-4500-4511
Apollo-11 6226-6220
Apollo-15 802-823 West, looking oblique
    1371-1391 East looking oblique
    1487-1504 North looking oblique
    0358-0378 (Condorset)
    0937-0957 (Auzont & Firemiscus)
    1620-1640 (Apollonius)
    Apollo-17 274-293 (North above 12'N to East above 14'N at West)
    Apollo-15, 16, 17 Panoramas
    S.W., Corner Mare Crisium, USGS I-722
    LO-IV-61, 191, 192-H
    Apollo-11 6230
    N.W – N.E. Half M. Crisium, USGS I-707
    LO-IV-H-61, 54, 177, 191, 192, 62, 55 and 67

BIBLIOGRAPHY

Bonnette, Dennis "Origin of the Human Species."

Burgess, F. and W. D. Whitley "Text Book of Hindu Astronomy."

Cortright, Edgar (1968) NASA SP-168 "Exploring Space with a Camera."
 NASA, Washington D.C.

DiPietro, Vincent and Gregory Molenaar (1982) "Unusual Martian Surface features."

El-Baz, Farouk and L.J. Kosofshy (1970) "The Moon as Viewed by Lunar Oribiter."
 NASA SP-200

Fielder, Gilbert (1965) "Lunar Geology." Lutterworth Press, London

Golden, Fred (April, 1985) "Discover"

Hoagland, Richard (1991 Fall Vol-1, No2) "Martian Horizons"

Hoagland, Richard (1991, Summer Vol-1) "Martian Horizons"

Hoagland, Richard (1992 Winter Vol-1, No.3) "Martian Horizons"

Hoagland, Richard (1994, Winter Vol-1, No.4) "Martian Horizons"

Hoagland, Richard (1995, Summer Vol-2, No.5) "Martian Horizons"

Kaysing, Bill "We Never Landed A Man on the Moon."
 and "We Never Went to the Moon."

Kopal, Zdenek (1960) "The Moon." Chapman and Hall, London

Kopal, Zdenek (1971) "A New Photographic Atlas of the Moon." Taplinger Pub. Co.

Leonard, George (1976) "Somebody Else is on the Moon" (David McKay, Pub.N.Y.)

Loomis, Alden (1965 Oct. Vol-76, No.10) "Some geological Problems of Mars."
 Geo. Soc. Of America Bulletin

Midnight (Feb 8, 1977)

Moore, Patrick (1953) "Guide to the Moon."

More, Patrick (1950) "The Planet Mars."

Mutch, Thomas (1972) "Geology of the Moon A Stratigraphic View."
 Princeton University Press, N.J.

NASA (1968) "The Mars Book." SP-179, 1968 Edition.

NASA (1974) "Mars as Viewed by Mariner 9." NASA SP-329
 NASA Washington D.C.

NASA SP-206 "Lunar Orbiter Photographic Atlas of the Moon."
 David Bowker and Kenrick Hughes (1971)

NASA SP-241 (1971) "Atlas and Gazetteer of the Near Side of the Moon."
 Guischewski, Kinsler and Whitaker

National Enquirer (Oct 25, 1977)

Nelson (1955) "There Is Life On Mars."

Schultz, Peter (1976) "Moon Morpohology" Univ. Tex. Press

Shneour, Elie and Eric Ottensen (1966) "Extraterrestrial Life: An Anthology and
 Bibliography" Nat. Acad. Of Sc.

Short, Nicholas (1975) "Planetary Geology." Prinston Hall, Inc.

Spurr, J. E. (1944) "Geology Applied Selenology." Science Press Printing Co.,
 Lancaster, Penn.

Spurr, J. E. (1949) "Geology Applied to Selenology." Volume - IV,
 Literary Licensing

Sreckling, Fred (1981) "We Descovered Alien Bases On The Moon."
 (G. A. F. International)

Surya Siddhanta

Technology and Youth (May 1968)

Thomas, Andrew (1971) "We Are Not The First" London, Sphere.

Wallace, Alfred Russell (1907) Is Mars Inhabitable?"

SOURCES AND LINKS

1.) FREE PDF of book: "Somebody Else is on the Moon." By George Leonard. https://ia600404.us.archive.org/16/items/SomebodyElseIsOnTheMoon/SomebodyElseIsOnTheMoon.pdf
2.) https://www.metabunk.org/threads/debunked-alien-base-on-the-moon-triangle-of-dots-photo-artifact.2965/
3.) http://www.themortonreport.com/discoveries/paranormal/aliens-on-the-moon/
4.) http://ufodigest.com/article/who-else-could-be-moon
5.) http://www.godlikeproductions.com/forum1/message218714/pg1
6.) http://www.paranormalnews.com/article.aspx?id=1185
7.) http://gizadeathstar.com/2011/05/the-idea-that-will-not-go-away-bases-on-the-moon/
8.) http://www.thescienceforum.com/pseudoscience/381-3-different-moon-images-same-location-different-objects.html#post527548

OTHER SOURCES LINKS:

Books, Tapes, Videos, DVD's documentaries: at
Weirdvideos.com   https://www.createspace.com/Preview/1143454

BOOK ORDER INFORMATION AND QUESTIONS:

c/o Ross S. Marshall P.O. Box 1191, Anacortes, Wa. 98221

ALIEN ARTIFACT SERIES:

Alien Artifacts, Volume-1

Sections 1-2 "Is Anyone Else on the Moon? The Search for Alien Artifacts According to George Leonard.

PREVIEW: https://www.createspace.com/Preview/1144263
ORDER HERE! http://www.amazon.com/dp/1495987760

www.ingramcontent.com/pod-product-compliance
Lightning Source LLC
Chambersburg PA
CBHW052248220526
45471CB00001B/247